T0291673

CAMBRIDGE LIBRARY COLLECTION

Books of enduring scholarly value

Technology

The focus of this series is engineering, broadly construed. It covers technological innovation from a range of periods and cultures, but centres on the technological achievements of the industrial era in the West, particularly in the nineteenth century, as understood by their contemporaries. Infrastructure is one major focus, covering the building of railways and canals, bridges and tunnels, land drainage, the laying of submarine cables, and the construction of docks and lighthouses. Other key topics include developments in industrial and manufacturing fields such as mining technology, the production of iron and steel, the use of steam power, and chemical processes such as photography and textile dyes.

Matthew Boulton

This 1939 work gives deserved recognition to the achievements of the engineer and businessman Matthew Boulton. Boulton's importance has generally been overshadowed by that of his partner James Watt, but he was a significant figure in his own right, particularly in relation to the Soho Foundry and his production of coins and medals. He belonged to a network of highly significant men of the period, including Josiah Wedgwood, Erasmus Darwin and Benjamin Franklin, and was a founding member of the Lunar Society of Birmingham. An engineer by profession, H.W. Dickinson researched widely, and published highly readable works on the history of the steam engine, Watt, and Trevithick, also reissued in this series. He succeeds in producing a work which appeals to the scientist, the historian and the general reader, without feeling obliged to over-simplify the technical details.

Cambridge University Press has long been a pioneer in the reissuing of out-of-print titles from its own backlist, producing digital reprints of books that are still sought after by scholars and students but could not be reprinted economically using traditional technology. The Cambridge Library Collection extends this activity to a wider range of books which are still of importance to researchers and professionals, either for the source material they contain, or as landmarks in the history of their academic discipline.

Drawing from the world-renowned collections in the Cambridge University Library, and guided by the advice of experts in each subject area, Cambridge University Press is using state-of-the-art scanning machines in its own Printing House to capture the content of each book selected for inclusion. The files are processed to give a consistently clear, crisp image, and the books finished to the high quality standard for which the Press is recognised around the world. The latest print-on-demand technology ensures that the books will remain available indefinitely, and that orders for single or multiple copies can quickly be supplied.

The Cambridge Library Collection will bring back to life books of enduring scholarly value (including out-of-copyright works originally issued by other publishers) across a wide range of disciplines in the humanities and social sciences and in science and technology.

Matthew Boulton

H. W. Dickinson

CAMBRIDGE UNIVERSITY PRESS

Cambridge, New York, Melbourne, Madrid, Cape Town, Singapore,
São Paolo, Delhi, Dubai, Tokyo, Mexico City

Published in the United States of America by Cambridge University Press, New York

www.cambridge.org
Information on this title: www.cambridge.org/9781108012249

This edition first published 1937
This digitally printed version 2010

ISBN 978-1-108-01224-9 Paperback

MATTHEW BOULTON

SOHO MANUFACTORY, *c.* 1781

Courtesy of the Assay Office, Birmingham

MATTHEW BOULTON

by

H. W. DICKINSON

Author of
'James Watt, Craftsman and Engineer',
'Robert Fulton, Engineer and Artist'
etc.

CAMBRIDGE
AT THE UNIVERSITY PRESS
1937

PRINTED IN GREAT BRITAIN

CONTENTS

List of Plates *page* vii

List of Figures in the text and Tailpieces viii

Preface ix

Chronicle of the life and works of Matthew Boulton xiii

Chapter I. Introductory 1

Trade and industry in Tudor times—Manufactures go north to the coalfields—Iron and hardware—Means of transport: pack animals, waggons, river navigations, canals—Rise of the Black Country—Inventions—Toy and buckle trades—Rise of Birmingham—Its inhabitants.

Chapter II. The Boulton Family 23

Rise of the family and its coming to Birmingham—Birth and education of Boulton—Enters his father's business—His inlaid buckle—Marriage—Death of his father and of his wife—The buckle-chape bill—Second marriage—Comes into a fortune—Friends and contemporaries.

Chapter III. Soho Manufactory 41

Origin of the word Soho—Partnership with Fothergill—Toys and buttons—Enticed to go abroad—Sheffield plate—Ormolu clocks—Royal patronage—Silver plate—Founds Birmingham Assay Office—Visitors to the Manufactory.

Chapter IV. Boulton and Steam Power 75

Extension of Manufactory—Shortage of water power—Boulton's steam engine—Watt's improvement of the steam engine—Roebuck's share—Watt visits Soho and meets Boulton—Watt's patent for the separate condenser—Partnership mooted—Roebuck bankrupt—Boulton acquires his share in the patent—Watt removes to Birmingham—Extension of patent—Partnership.

Chapter V. Boulton and Watt 89

First engine in Midlands—Engines in Cornwall—Engages William Murdock—Keir's sheathing metal—Mechanical paintings—Letter copying—Birmingham Metal Company—Death of Fothergill.

Chapter VI. The Rotative Steam Engine 113

Watt's patents—Boulton's inventiveness—Death of his second wife—Breakdown in health—Visits Ireland and Scotland—Albion Mill—Foreign privileges—Centrifugal governor—Darwin's panegyric—Argand's lamp—Pitt's taxation—Cornish Metal Company.

Chapter VII. Coinage and Soho Mint *page* 133

State of the coinage—Counterfeiting—Boulton's ideas—Appears before the Privy Council—Applies steam power to coinage—Patents coining press—Erects Soho Mint—Work as a medallist—Coinage of Great Britain—Fits out Royal and other Mints—Copper sheathing for ships.

Chapter VIII. Soho Foundry 163

Change in manufacture—Education of partners' sons—Taken into partnership—New works—Rearing feast—Patent litigation—Prices and performance of engines—End of partnership of Boulton and Watt—Soho Insurance Society—Soho House.

Chapter IX. Declining Years 186

Lunar Society—Illness—Death—Funeral—Character—Appearance—Portraits—Distinctions—Will—Epilogue.

Appendix I. Memoir of Boulton by Watt, 1809 203

Appendix II. Boulton's Businesses 209

Index 210

LIST OF PLATES

Soho Manufactory, *c.* 1781 *frontispiece*

I. Some of Boulton's friends and associates *facing page* 38

II. Wedgwood cameos in Boulton cut steel frames 46

III. Cut steel fob chains 47

IV. Sheffield plate candelabrum, *c.* 1800 53

V. Ormolu and Blue John candelabrum at Windsor Castle, *c.* 1770 57

VI. Silver tureen, 1776–77 70

VII. The "Lap" engine, 1788, showing the "sun and planet" gear 125

VIII. Matthew Boulton, *aet.* 64 132

IX. The art of coining in 1750 134

X. Some of Boulton's coins and medals, 1788–1805 158

XI. Soho Foundry, *c.* 1820 172

XII. Original entrance to Soho Foundry 173

XIII. Engraving at the head of the Soho Insurance Society's Rules, 1792 180

XIV. Soho House, present day 184

LIST OF FIGURES IN THE TEXT

1. Part plan of the township of Handsworth, 1794 *facing page* 42

2. Diagram of Newcomen's steam engine, 1712 *page* 77

3. General section of Watt's engine, 1776 91

4. Boulton's return tube steam boiler, 1781 118

5. Boulton's coining press, 1790 141

6. Map of distribution of Boulton and Watt engines, 1775–1800 177

TAILPIECES

Boulton Memorial Medal, 1819, obverse 162

Boulton Memorial Medal, 1819, reverse 202

PREFACE

It has long been a matter for surprise that no separate biography of Matthew Boulton has been written. We are obliged to lay stress on the word "separate" because Smiles in his *Boulton and Watt*, 1865, has treated faithfully the work of Matthew Boulton and of James Watt in conjunction, particularly in their improvements of the Steam Engine. Naturally, since fresh facts have come to light with the lapse of time, some of Smiles's conclusions have now to be restated.

The gap in our biographical literature became very apparent when, last year on the occasion of the bicentenary of Watt, attention was directed anew to his life and achievements.[1] Even to the superficial student of the facts of Watt's life it was obvious that, without Boulton's enthusiasm and enterprise, Watt could hardly have succeeded in bringing his improved steam engine before the world, and further it was clear that Boulton created the environment in which Watt was enabled to work out those further improvements in the steam engine that transformed it from an apparatus merely for lifting water into one with an immensely larger field of application to general power purposes. These improvements were so great that they amounted to a new birth of the engine. The part played by Boulton in this development has been too much overshadowed by that played by Watt and this overshadowing has been unfortunate because, had Boulton taken no part in the introduction of the steam engine, he would still deserve a niche in the temple of fame for his many other achievements.

First and foremost he was the most prominent citizen of Birmingham during the eighteenth century—a period of unexampled development; his public-spirited actions did much to raise the status of the town by helping to remove the stigma under which its products laboured; he brought in new industries

[1] Dickinson, H. W., *James Watt, Craftsman and Engineer*, 1935.

and enlarged old ones; and he was largely responsible for the introduction of the new system of industry known as the factory system. Many of his productions were of a high artistic merit and are only now beginning to be recognized and appreciated at their true worth.

Perhaps this very devotion to the interests of his native town inhibited a wider reputation in the national and political sphere such as with his great gifts would have been within his reach—such a reputation in fact as we have witnessed in the career of a fellow-manufacturer from Birmingham, the centenary of whose birth has been celebrated this year.

Considerations of this order induced the author to attempt to fill this gap in the lives of our great men. He found quickly that the aspects of Boulton's life were so many that to do justice to them no single author could hope to command the qualifications that are necessary to enable him to do so. Indeed it has been borne in upon the author that this may be the reason why the story of Boulton's life has not previously been attempted. The author is conscious that his training, being that of an engineer, has led him to lay stress upon Boulton's technical rather than upon his artistic achievements; the author pleads for leniency wherever his treatment is inadequate. He feels also that, being neither Birmingham born nor bred, he is exhibiting temerity in rushing in where others have feared to tread. However, the temerity is more apparent than real, for within the limits of space imposed, an exhaustive and fully documented biography is not possible. That remains as a task to be undertaken by some future writer; the present work must be looked upon only as a start in the right direction. It is enough to say that the materials available are abundant and of absorbing interest, not only in relation to Boulton but also to the industrial and technical history of the eighteenth century.

The most important store of documentary material is in the Boulton Papers, formerly in the archives at Tew Park, Oxon, the present family seat, but in 1926 confided to the care of the Assay Office, Birmingham, itself, as we shall learn later, an institution that owed its foundation largely to Boulton. To these materials the author has been given access in the most generous manner through the kindness of Mr Arthur Westwood, the Assay Master. He it was that the author always hoped would become Boulton's biographer for no one has a more intimate knowledge of his life and no one is more willing to impart that knowledge. It was only when the author was satisfied that Mr Westwood's official duties did not afford him time for the task that the present work was undertaken. The material at the Assay Office is so extensive and is so largely used in the succeeding pages that, where no reference to source is given, the reader is asked to assume that it is from the Boulton Papers.

Almost as large an amount of material, but relating to the firm of Boulton and Watt only, is preserved in the Public Reference Library at Birmingham where it is known as the Boulton and Watt Collection (referred to as "B. and W. Coll."). The author is indebted to Mr H. M. Cashmore, the Chief Librarian, and to his efficient staff, not only for help in supplying material from the Library, but also for help in hunting down information. The assistance of Mr C. S. Kaines of the Birmingham Art Gallery is also acknowledged. The author is pleased to have had the approbation of, and help in his task from, descendants of Boulton now of the third and even of the fourth generation. The family tree that will be found on p. 27 owes its completeness to this help and to that of the Rev. C. E. Salisbury, the Vicar of Great Tew.

Research has been prosecuted at the Royal Palaces, the Record Office, the Science Museum, the Victoria and Albert Museum, the Libraries of the Patent Office and of the British Museum, the Royal Institution and at the Heralds' College. To the Officers of these Institutions, many of them friends and former colleagues, the author expresses his warmest thanks for their courtesy and help. The author's son, H. D. Dickinson, M.A., has been of great assistance, especially in the introductory chapter. Members of the Newcomen Society and other friends have been active in supplying information. Had space served, the author would have liked to have given the names of these gentlemen with an enumeration of their good deeds. Lastly the resources of the Cambridge University Press have been placed at the disposition of the author in such wise as would lead him to imagine, if he did not know better, that his pebble was the only one on the beach.

H. W. D.

Purley, Surrey
25th July 1936

CHRONICLE

of the

LIFE AND WORKS OF MATTHEW BOULTON

Date	Age	
1728	—	Born September 3rd (September 14th New Style) at Birmingham.
1745	17	Enters business of his father.
—	—	Invents inlaid steel buckle.
—	—	Marries Mary, d. of Luke Robinson, of Lichfield.
1759	31	Death of his father. Death of his wife.
1762	34	Builds Soho Manufactory and enters into partnership with John Fothergill.
1766	38	Builds Soho House. Co-founder of Lunar Society.
1767	39	Extension of Manufactory. George III patronizes the products of Soho. Marries Anne, d. of Luke Robinson, of Lichfield.
1768	40	First meets James Watt.
1772	44	Affected by trade depression.
1773	45	Prime mover in the establishment of the Assay Office in Birmingham.
1774	46	Acquires Dr John Roebuck's interest in Watt's steam engine patent. Watt comes to Birmingham.
1775	47	Extension of Watt's patent for 25 years. Enters into partnership with him for like term.
1777	49	Engages William Murdock.
1778	50	Pays his first visit to Cornwall. Engages Francis Eginton to copy pictures.
1779	51	Again visits Cornwall. Enters into partnership with Watt and Keir in copying press business.
1780	52	Becomes interested in the Cornish mines.
1781	53	Dissolves partnership with Fothergill.
1782	54	Death of Fothergill. Enters into partnership with John Scale.
1783	55	Death of his second wife. Breakdown in health. Visits Scotland. Starts Albion Mill.
1784	56	Elected Fellow of the Royal Society of Edinburgh. Joins in loyal address to George III. Opposes Pitt's taxes on raw materials. Founds Chamber of Manufacturers.
1785	57	Helps to found the Cornish Metal Company. Elected Fellow of the Royal Society, London.
1786	58	Applies the steam engine to coining.
1788	—	Appears before the Privy Council on the coinage question.
1789	61	Rebuilds Soho House.

Date	Age	
1790	62	Takes out patent for coining press.
1791	63	Church and King riots in Birmingham. Soho threatened.
1794	66	Firm of Boulton, Watt and Sons founded.
1795	67	Soho Foundry built. High Sheriff of Staffordshire.
1797	69	Employed in the coinage of Great Britain.
1798	70	Patents the hydraulic ram for Montgolfier.
1799	71	Begins to fit up the Royal Mint with his coining machinery.
1800	72	Partnership with Watt ended. Firm of Boulton, Watt and Co. founded.
1807	79	Helps to establish theatre in Birmingham. Attacked by illness.
1809	81	Dies at Soho, August 17th, and is buried in Handsworth Church.

CHAPTER I

INTRODUCTORY

Trade and industry in Tudor times—Manufactures go north to the coal-fields—Iron and hardware—Means of transport: pack animals, waggons, river navigations, canals—Rise of the Black Country—Inventions—Toy and buckle trades—Rise of Birmingham—Its inhabitants.

THE life of Matthew Boulton, the subject of the present memoir, was bound up with the town of Birmingham and in turn influenced greatly its growth and development during the eighteenth century. He was among the earliest of its townsmen to achieve a more than local reputation, and in doing so helped the name and products of the town to become known to the ends of the earth.

To understand how the stage was set that made this possible, something must be said about the rise of the Midlands and of Birmingham in particular to the position it had attained at the beginning of the eighteenth century, although within the limits of space imposed upon this volume only a brief sketch is possible.

For our purpose it is not necessary to look further back than the Tudor period, commencing, say, in 1500. The policy then inaugurated by the advisers of the Crown and steadily pursued subsequently was to advance from the existing agricultural economy to one of manufacturing industry. This is an explicit statement of what the policy in effect amounted to rather than what was consciously in view. As a matter of fact, we find that a great part of the energies of the government was devoted to the maintenance of the *status quo* in agriculture. The underlying object of policy was the attainment of military power; industrial progress was a by-product of the pursuit of this policy. Industry was fostered because it promoted military ends, both directly and indirectly: directly, insomuch as it provided the sinews of war: shipbuilding, rope-making, brass-founding, cutlery and saltpetre monopoly; indirectly, through trade, for in those days of primitive banking when no such thing as a public loan was possible, every prince tried to amass as large an amount as possible of gold and valuables as a reserve for war. If he had no mines of precious metals within his dominions, he encouraged a favourable balance of trade, so that bullion should come into his realm in payment for an excess of exports over imports. This was achieved negatively by discouraging imports through high tariffs, licences, prohibitions, etc., and positively by the encouragement of exports. Now in those days, the only goods that could be traded in overseas were those that had great value in little bulk embodied in them, either precious raw materials or highly-skilled labour—in other words every state tried to stimulate the more highly-skilled branches of manufacture with which to establish an export trade.

Another reason why state policy fostered manufacture and foreign trade was convenience in levying taxation. Agricultural wealth consisted largely of goods which were destined for immediate consumption, goods which in any case could not be realized as money values owing to the absence of a market or the virtual absence of circulating capital in the hands of landlords and farmers. Mercantile wealth, however, was easily realizable, and therefore could be assessed and taxed in terms of money, especially in their passage across frontiers. In a community of the sixteenth-century stage of development, indirect taxation is much more convenient than direct taxation.

Thus an alliance grew up between the state power and the mercantile interests; the latter received favours from the former (e.g. tariffs, monopolies, charters for companies) and in return provided the former with the sinews of war through taxation on trade.

This new policy was carried into effect by fostering existing industries, by bringing new manufactures into the country; by encouraging the settlement of artisans therein and by the grant to them of privileges or patents of monopoly. This change in the economy of this country—and the change was a world-wide one, although perhaps not so early in other countries as in this—was accompanied by a change from a way of life based almost purely on animal and vegetable products to one based extensively on mineral products in addition to the other two. No longer could the surface of the earth support man's needs—he now sought underground the riches that could add to his convenience and comfort. Hitherto power had been generated by muscular effort—the harnessing of animals, or by impounding water to turn millwork, or by taking advantage of the wind in a windmill; fuel for warmth or industrial purposes came from the woodlands and if for the last-named purpose was usually in the form of charcoal; salt was evaporated from sea water; lye or potash was lixiviated from wood ashes to make soap with animal fat and so on. It is true that metals such as iron had long previously been extracted from their ores but this had been done only in small quantities.

By the sixteenth century, the exploitation of purely surface products, in carrying out the new policy, had reached its limits. The product, the need for which made recourse to a mineral economy inevitable, next to metals, was fossil fuel. This had been used already for centuries locally and spasmodically as a domestic fuel, in London, for example, whither coal was brought from Newcastle-upon-Tyne. Since it was borne coastwise it received its common appellation of "sea coal". Its use for industrial as well as for domestic purposes now began.

It is well to fix in the mind that this far-reaching change was really one, and perhaps the most important one, of those changes which led on later to what has received the somewhat facile designation of the industrial revolution. If the subject from the Tudor period to the present day is viewed as a whole it will be found that one industry after another underwent, in either its technique or its conditions of labour or in both, similar changes, violent for the time being, and then settled down again. Not only so, but these violent changes were not single phenomena even in one industry; in the iron industry, for example, at least four such periods can be traced. Nor did the changes affect only the industry concerned; they had their ramifications throughout the industrial fabric. Further the process is still going on and it is no prophecy to say that it will continue. If a graph could be prepared on which the duration and amplitude of changes in a number of the most important industries could be represented and superposed, we should get summations which would explain why certain periods in our industrial history have been so fateful.

But to resume the thread of the story. It is usually stated that the change over from vegetable to mineral fuel was caused by the scarcity of wood and the demands upon timber for shipbuilding. The various Acts of Parliament (Hen. VIII, 1543; Elizabeth, 1558, 1581) restraining the use of the woodlands for making charcoal, lend support to this view, but it will be found that timber trees suitable for shipbuilding were not used to make charcoal. Wood for this purpose in the form of coppice, cut down every 15 or 21 years, was grown as a crop just as much as say cereals. It was this coppice wood along with the smaller branches of the timber trees (when cut down at very much longer intervals) that were used for conversion into charcoal. These coppice woods could have continued indefinitely to supply industry—not perhaps on a much larger scale than then practised—but the real reason for the change was an economic one.[1] The gathering ground for the crop is a large area. The cost of "coaling" wood is very high, the product is bulky and the cost of transport is a heavy item. In fact once the technique of using the much more concentrated and easily won fossil fuel had been mastered, for the industry concerned, there was no question of using charcoal any longer. The change over took place in one industry after another as rapidly as prejudice and economic considerations would allow.

The need for cheap fuel was felt earliest perhaps in the manufacture of iron because such a large quantity of fuel is required—in the case of charcoal about 20 to 1 of the iron produced. We are to remember too that iron was still produced by the age-old process of direct reduction from the ore in the bloomery in quantities of a few pounds weight. Hence we are to think of iron as a rare commodity, so much so that in the Middle Ages it constituted easily the largest single item of cost in agriculture.

As early as the fourteenth century, however, the smith had

[1] Cf. Straker, E., *Wealden Iron*, 1931, p. 109.

learnt to use mineral fuel for working up the bar iron, once it had been produced in the bloomery, into implements and tools required in husbandry and trade. At first, probably, the fuel was used in admixture with charcoal, and perhaps with peat, but later the smith learnt, by a kind of coking process on his own hearth, to drive off much of the volatile matter and the deleterious sulphur from the coal, thus obtaining a fuel like that we know to-day as smithy char.

This gave an impetus to the working up of iron for other than agricultural purposes, particularly nail making. Hence we find that such trades, designated commonly as hardware, were attracted or migrated to the coalfields, especially those of Newcastle, Warwickshire and South Staffordshire, which were the first discovered and worked. It was in the last-named area that the celebrated 10-yard seam of coal was found.

The main reason why the hardware trades were able to expand was because iron became more plentiful owing to the introduction from the Continent *c.* 1500 of the blast furnace which yielded cast or pig iron in large masses as compared with the insignificant output of the bloomery. This pig iron was converted into wrought iron by the refinery process. It is to be emphasized continually that up to the stage of wrought or bar iron, the fuel used was still exclusively charcoal. The technique of using mineral fuel in the blast furnace was not mastered till the first decade of the eighteenth century by the celebrated Abraham Darby, of Coalbrookdale, who found the way to coke the coal before using it in the furnace. Even then for many years his product was not found suitable for refining into wrought iron but was worked up in the form of castings for which a vast field of use developed. The effect on industry of these changes was enormous and we shall have a word to say about it later.

It was but natural that, when manufactures were being introduced or fostered in this country under the new policy, they should be seated largely in the South of England on the sea board or on estuaries, for that is where their products were wanted; thus we find pig-iron making in the Weald of Sussex, glass making in Surrey, paper making in Kent, shipbuilding on the Thames, woollen manufactures in Berkshire, Norfolk and the West of England and so on. In nearly every one of these industries power was required and the only way to obtain it was by the muscular effort of animals, by wind-mills or by impounding water to turn water-wheels. Animal-power was extensively used, e.g. in the horse-mill, but was limited in scope and was costly. The wind is so fitful that its only general application was to corn milling. Water-power was most generally convenient. Now in the Midlands and the North of England water-power is more reliable and constant than in the South because the rainfall is higher and the configuration of the land is more favourable in the former areas than in the latter. Thus the migration of industry northward had begun even before coal had become a determining factor. It is passing strange that we are experiencing to-day a reversal of direction, and industries are coming south again, but the cause of this is outside our present ambit.

The old economy of the production of goods on the spot where they were to be consumed having thus given way to trade on the large scale, improved means of transport both by land and water became necessary—indeed insistent. In the country the network of roads left by the Romans had been allowed to fall into such an incredibly shocking state by the seventeenth century that the roads could be said to be impassable for wheeled carriages in winter and little better in summer. Such transport of goods as was required was effected by the old method of pack animals but naturally it was slow, costly and limited as regards the weight that could be carried. The ramifications of the system were remarkably far-flung and many a green track and occasionally a packhorse bridge, such a bridge as the Essex Bridge over the Trent at Great Haywood, Staffordshire, or the bridge over the Derwent in Derbyshire soon to be moved when the new Derwent Reservoir is completed, remain to bear witness to the extent of pack transport.

The bad state of the roads led to increased use of water transport. Hence coasting trade in vessels of about 100 tons burden sprang up, bringing in its train the construction of quays, harbours and docks. Estuaries such as the Thames and Severn were the scenes of the earliest activity. For example, Pepys mentions in 1661 that a dock was being dug at Blackwall, and in 1665 he records a project for making a dry dock at Woolwich. A concomitant of water-borne traffic is shipbuilding with its ancillary trades of rope, canvas, chain and anchor making and of saw milling. Shipbuilding yards grew up wherever there was a suitable "hard", that is an unyielding gravel beach shelving quickly into deep water, preferably in the vicinity of a trading town; we may instance Deptford Strond that gave birth among other matters of moment to the Corporation of the Trinity House. When as early as the reign of Henry VIII, vessels began to be differentiated according as they were to be used for trading or for naval warfare, Royal Dockyards were established.

As the coasting trade developed attention was directed more and more to the improvement of the rivers by deepening the beds, by cutting new channels and by constructing "flashes" or locks; the resulting works became known as navigations. Between the years 1660 and 1670 at least ten such projects were approved by Parliament among which we may cite as pertinent to the present enquiry the Worcestershire Stour and Salwarpe Navigation, 1662. Incidentally it may be noted that as early as 1635 the Severn was considered to be navigable as far as Shrewsbury.

A second outburst of activity in river navigation occurred between 1697 and 1700. Among the projects carried out in this period was the deepening of the Warwickshire Avon between Tewkesbury and Stratford, 1700.

The speculative mania that swept the country, best exemplified in the South Sea Bubble, did not pass without affecting navigations; between 1719 and 1721 some nine Acts of Parliament approving such schemes were passed.[1] It is true that lack of capital hindered or curtailed much of this activity; nevertheless there grew up a race of undertakers or engineers capable of directing such works, and another race of navigators or navvies capable of executing them.

[1] Cf. Willan, T. S., *River Navigation in England*, 1600–1750, 1936.

Nor were canals or purely artificial navigations unknown in this country. A Thames-Severn canal was projected as far back as 1606 by James Jessop, and by others, as was a Thames-Bristol-Avon canal about the same time by Henry Briggs and brought forward again by Francis Matthew in 1655. These were, it is true, paper schemes but they exhibited the spirit of the time and presaged the future—a network of inland waterways joining the navigable rivers to the seas and constituting a complete system of water transport. This third and closing link in the system did not receive the attention it deserved until the time that the Duke of Bridgewater's Canal at Worsley, near Manchester, engineered by James Brindley and executed in 1767 in the teeth of great difficulties, had attained success.

A period of great activity in canal building ensued, culminating about 1795 in a positive canal mania. It is but just to say that all the elements of canal construction and operation—pound-locks, reservoirs, inclined planes and barges—were known on the Continent and that no invention in connection with canals appears to have been of English origin, but the steady growth of skill, knowledge and experience gained in river navigation that preceded the canal era, and made its realization possible, has been too much overlooked.

With these observations we may now narrow down our enquiry to the area of the South Staffordshire Coalfield or, as it came to be called, the Black Country. We know little about it before the fourteenth century for till that time it was mostly waste land sparsely populated and, owing to its situation away from roads and rivers, as remote as can be imagined from the outside world. The district was, however, endowed by nature with supplies of timber, ironstone, coal and other raw materials. Charcoal made from the timber supplied fuel for the reduction of the ironstone in the bloomery; the bark of the oak trees supplied the tan pits with liquor to tan leather; coal first dug here in the fourteenth century began to supplement charcoal as fuel for the smith. In the following century a gradual increase took place in the population, engaged as colliers (i.e. charcoal burners) and smiths living in widely scattered small communities;

this led to the beginnings of trade in their products with the larger world outside, necessitating the thrusting out of pack-horse tracks and the making of roads.

The blast furnace introduced from the Continent into Sussex early in the sixteenth century did not reach the Midlands till about the end of that period. The technique of the smith had meanwhile developed and differentiated into the distinct trades of nailsmith, chainsmith, locksmith, bladesmith and cutler. For the production of edge tools—axes, adzes, sickles, scythes; boring tools—bradawls, gimlets; and cutlery—knives—it was necessary to "steel" the cutting edge of the tool, i.e. to weld a thin piece of steely iron to the wrought-iron backing. This kind of steel was produced aberrantly in the bloomery and the experienced bloomsmith looked out for it and reserved it for making tools. Naturally the material varied much in hardness and the resulting tool made from it in quality.

The first writer to throw light upon what was happening in the Black Country was John Leland or Leyland, the traveller and antiquary, who visited these parts about 1538; he described what he saw and learnt in his *Itinerary*. His description has been quoted times out of mind but it is so vivid and informative that we must be pardoned for reproducing it.[1] Speaking of "Commodities of the Soile", he says:

Se Coles at Weddesbyri (Wednesbury) a village a 5 miles from Lichefelde by West South West.

Waulleshal (Walsall) a little Market Toune in *Stafordshir* a mile by North from Weddesbyri. Ther be many Smithes and Bytte-Makers yn the Towne. It longgith now to the King and here is a Parke of that Name scant half a Mile from the Town yn the way to *Wolverhampton*.

At Walleshaul be Pittes of Se Coles, Pittes of Lyme that serve also Smith Toun 4 miles of. There is also Yren Oare.

Further on[2] Leland speaks of Birmingham in these terms:

I came through a pretty Street or ever I entred into *Bermingham Towne*. This Street, as I remember, is called *Dirtey*. In it dwell

[1] *Itinerary*, edition 1711, vol. VII, 29.

[2] *Loc. cit.* vol. IV, p. 89.

Smithes and Cutlers, and there is a Brooke that divideth the Street from *Bermigham* and is a Hamlett or Member belonginge to the Parish therebye.

This Brooke above *Dirtey* brancheth in 2. Arms that a little beneath the Bridge close againe This Brooke riseth, as some saye, 4. or 5. Miles above *Bermingham* towardes *Black Hills.*

The Beauty of *Bermingham*, a good Markett Towne in the extreame parts of *Warwickeshire* is one Street going up alonge almost from the left Ripe of the Brooke up a meane Hill by the length of a Quarter of a Mile. I saw but one Paroch Church in the Towne. There be many smithes in the Towne that use to make Knives and all mannour of cutting Tooles, and many Loriners that make Bittes, and a great many Naylors, Soe that a great part of the Towne is maintained by Smithes, whoe have theire Iron and Sea-Cole out of *Stafford-shire.*

There is at the end of *Dirtey* a proper Chappell and Mansion House of Tymber, hard on the Ripe as the Brooke runneth downe, and as I went through the Ford by the Bridge, the Water ranne downe on the right Hand, and a few Miles lower goeth into *Tame ripa dextra*....

Leland may not have familiarized himself with the pronunciation of the inhabitants, or his memory may not have served him too well, for the name of the street was not Dirtey but Deritend (now High Street, Deritend). The "chappell" was that of St John, Deritend, founded in 1375. Ripe, *Lat.* ripa, means the river bank. The ford was that across the brook Rea, a tributary of the river Tame.

Half a century later, i.e. in 1586, we have the account of William Camden, the antiquary, which does not differ in any material respect from that of Leland. Camden says[1] (translation) that six miles from "Sutton Colfield" is "Bremicham swarming with inhabitants and echoing with the noise of anvils for here are many blacksmiths". Other topographers write in a similar strain, e.g. in 1627: "Bremincham, inhabited with blacksmiths and forging sundry kinds of iron utensils."

During the seventeenth century the expansion both of trade and population in the district was very great. Perhaps the trade

[1] *Britannia.* 1586. His words are "Sex ad occasum hunc millioribus disjungitur Bremicham, incolis repertum & incudibus resonans sunt enim plurimi fabri ferrarij."

of nail making is the key to this rise. The trade had been long established in a small way at Stourbridge. A great impetus to it was given by the introduction there about 1628 by Richard Foley (1580–1657) of the slitting-mill for making nail rods. The mill used by him consisted of two rollers supported in heavy bearings or housings; the rollers had narrow collars meshing together such that on passing a sheet of hammered iron between them, it was sheared or slit into strips the width of a collar, i.e. into rods of such dimensions that nails could be quickly forged from them. Prior to this invention, nail rods were made by the laborious and tedious method of cutting up the hammered sheet when hot by a cold sett or chisel and sledge hammer, just as the smith cuts up a piece of iron to-day.

Many romantic stories are told of the introduction of new processes or inventions and none is more moving than the tale of the introduction of the slitting-mill. We are told how Richard Foley went to Sweden, where the mill was in use, and in the disguise of a fiddler got access to works where iron was slit. He brought away, as he thought, the secret but after his return to England found that there were details he had not mastered. He went to Sweden a second time to the same place and so delighted were his quondam associates at his return that they lodged him for safe custody in the mill itself. Foley was thus enabled to study every detail at his leisure. Alas for this delightful and circumstantial story! The prosaic fact is that the slitting-mill had been introduced into this country in 1599 from Liége by one Godfrey Box who had been encouraged by the grant of a patent. He established a mill at Dartford in Kent and this was still working thirty years after its introduction. We have nothing definite to prove that Foley saw this mill or that he came to any arrangement with the patentee—it is significant that the patent had expired by then—but obviously Foley did not need to go outside this country to obtain all the information he needed. Foley was most successful and amassed a fortune.

In corroboration of the fact that expansion in the nail-making trade did take place we have the evidence of Dud

Dudley (1599–1684), not over-reliable as to matters of fact, however, who, writing in 1665, computed that there were "twenty thousand Smiths or Naylours at the least dwelling near these parts", i.e. the district around Dudley. Even if he were exaggerating, the actual number must have been great.

It is hardly possible to stress too strongly the importance of inventions like the slitting mill in promoting the hardware industry, and the readiness with which they were adopted in this country, as the following quotation shows:

> FEW Countries are equal, perhaps none excel the *English* in the Numbers and Contrivance of their Machines to abridge Labour. Indeed the *Dutch* are superior to them in the Use and Application of Wind-Mills for sawing Timber, expressing Oil, making Paper and the like. But in regard to Mines and Metals of all Sorts; the English are uncommonly dexterous in their Contrivance of the mechanic Powers; some being calculated for landing the Ores out of the Pits, such as Cranes and Horse-Engines:—Others for draining off superfluous Water, such as Water Wheels and Steam Engines: Others again for easing the Expence of Carriage such as Machines to run on inclined Planes, or Roads down Hill with wooden frames, in order to carry many tons of Materials at a Time. And to these must be added the various Sorts of Levers used in different Processes: Also the Brass Battery Works, the Slitting Mills, Plate, and Flatting Mills, and those for making Wire of different Fineness. Yet all these, curious as they may seem, are little more than Preparations or Introductions for further Operations. Therefore when we still consider, that at *Birmingham*, *Woolverhampton*, *Sheffield*, and other manufacturing places, almost every Master Manufacturer hath a new Invention of his own, and is daily improving on those of others, we may aver with some confidence that those parts of *England* in which these things are to be seen, exhibit a Specimen of practical mechanics scarce to be paralleled in any part of the World.[1]

It is pertinent therefore to mention one or two more inventions of the kind described above. Such a one was that of cementation steel, that is steel produced by baking wrought iron embedded in powdered charcoal in chests at a red heat for four or five days; in the atmosphere so created the wrought iron becomes impregnated with carbon or converted into steel, for

[1] Tucker, Josiah, *Instructions for Travellers* (ed. 1757), p. 20.

the steel of that day differed from wrought iron only in the amount of carbon it contained. By this process steel of any desired content of carbon or hardness could be obtained at will. Reliable tools could be made from it. The process was the invention of William Ellyot and Mathias Meysey who patented it in 1614. It is not known how soon thereafter the technique reached Staffordshire but it is on record that by 1686 John Heydon had established a cementation furnace at Kingswinford.

Another striking invention—the adjective is used literally—that greatly helped the hardware trades, was that of the screw press or fly-press—a simple tool of amazing range. Screw presses were in use for stamping buttons, etc. in London and other parts of the country in the latter half of the seventeenth century. By Act of Parliament, 1696, the possession of a press which might be used for coining was made illegal. This was repealed early in the eighteenth century. The press consists essentially of an open frame supporting a nut in which works a vertical quick-threaded screw, to the top of which is fixed a lever or handle heavily loaded. The screw is vertically over the bolster in the bed plate, supporting the die, while its counterpart is on the end of the screw. By giving the loaded lever a smart pull, the upper die can be brought down on the lower one with a momentum that at the instant of striking is so great as to force a thin piece of metal interposed between the dies into practically any desired shape. Our illustration, Pl. IX, shows the fly-press as used in 1750 for coining, and its mode of operation on heavy work such as this. Making the screw was the chief difficulty because there was no method of doing this except cutting it laboriously by hand. The resulting screw could not be accurate, but having cut it, the nut was cast on it. However, the accuracy of the press did not depend on that of the screw because the die was not rigidly fixed to the screw but moved in guides at the moment of impact. The unlimited capacity of the tool in cutting out thin metal and in forcing it into any desired shape, its rapid operation and its cheapness for repetition work, opened out new fields for manufacture.

We have perhaps stressed over duly the importance of iron

making in the economy of the Staffordshire area. There were other industries that rose to importance. Copper was not mined in England before the sixteenth century, but one of the producing areas was in North Staffordshire at Cheadle where it was worked up; an extensive brass manufacture also grew up. Brass was obtained by melting copper with calamine, an ore of zinc. Both copper and brass were valuable raw materials for Birmingham industries. At Stourbridge, the occurrence of white sand and the existence of refractory clay above the coal measures attracted glass makers to settle there early in the seventeenth century. These workers had been brought over earlier still from French Lorraine and had been settled in Sussex and elsewhere. The Stourbridge sand, with alkalies furnished by wood ashes and with other ingredients, was melted in crucibles of the local clay by the aid of the coal on the spot. The glass so made was used for the production of bottles, drinking glasses and window glass.

The tendency of a trade to develop in a particular area and to be segregated there is very noticeable in the Black Country. Lock-making has been a special feature at Wolverhampton and Willenhall; saddlery and leather goods at Walsall, where later brass and copper wares have settled: edge tools at Cannock; chain-making and anchor smithing at Cradley Heath and so on.

Then again industries have fluctuated and have migrated within the area itself. Nail making, that was so important a trade in Birmingham in Leland's day, by the seventeenth century had shifted westwards only to be concentrated in Dudley and in Stourbridge where it had been settled originally. Edge tools, knives and swords were made in Birmingham in the seventeenth century but removed to other towns or left altogether. The lorimers departed to Walsall quite appropriately to join the saddlers there. The places of these trades were taken by others brought in from outside—a practice that has ever been characteristic of the enterprising inhabitants of the district. A good example of this is the making of cast-iron hollow-ware—pans and kettles—which became a Walsall industry not so many years after the method of making the ware had been introduced into this country from the Netherlands. To cast a thin-walled hollow article moulded in dry sand was a difficult operation but Abraham Darby succeeded in doing it, obtained a patent for it in 1707, and practised the process at Coalbrookdale, Salop, for many years with closed doors. It is of some interest to know that the first patent granted to an inhabitant of Birmingham, Richard Baddeley, ironmonger, 1722, Mar. 22, was for a hollow article in cast iron, viz. a box smoothing iron for linen; the patent included also cast iron "strakes" for cart wheels.

The trades that took the place of those that migrated were ever more and more highly-skilled, particularly in Birmingham. Hence we find the button, buckle and "toy" trades introduced there late in the seventeenth century. The button trade was brought in about 1660 partly owing to a tariff policy and partly owing to the falling-off in trade because of war with France; a further stimulus was given to it by the general growth of luxury and by changes in fashion. Samuel Pepys, always delighted when he could command the latest sartorial habiliments, mentions with satisfaction gilt and silver buttons on his coats. We shall have more to say on this head. Other trades may rise and fall but the trade in buttons goes on and on and seems destined to do so as long as western civilization endures. The torrent of buttons from Birmingham has now continued for over two centuries and saturation point, it might be thought, should be within reach, what with new forms of fasteners and in spite of the constant search and fashion for novelties in such a simple article. But no! buttons, like pins, disappear and leave no trace.

Buckles are in a different category. They were first introduced into England in the time of William and Mary, about 1690 or before, and fashion decreed that every man should wear them on his shoes, on his knee-breeches and even on his stock. The trade was taken up originally at Walsall but soon migrated to Birmingham where it grew to enormous dimensions helped by the numberless varieties in material, size and style dictated by fashion from day to day. The first note of decay in the use of the shoe-buckle occurred in 1786 when the shoe-lace began to be substituted. In the army the buckle had to be abolished because it tended to foul the stirrup and thus became a danger. The change in dress, dispensing with buckles, came about in the first decade of the nineteenth century.

Of toys we shall give details later.

The impression left on the mind by a study of the South Staffordshire area is, firstly, rapid development of natural resources; secondly, the conversion of products, native and imported, into an ever-growing variety of articles, associated with an ever advancing standard of handicraft, leading to specialization and to segregation of trades in different areas; thirdly, to the growth of population not only of craftsmen but also of enterprising men, restless in the search for new industries, alert to the march of invention and quick to realize their opportunities and to profit by changes of fashion.

We have already touched upon the question of transport in England generally; it is now desirable to form some idea of how the particular needs of the South Staffordshire area in this respect have been met, not only for the marketing of the finished goods but also for the importation of the hundred and one materials from all parts of the world used in the industries of the area.

Even in respect of wrought iron, the one raw material that it might be thought there would be no necessity to import, quite a considerable trade in Swedish and New England iron grew up, because these two kinds were found to be those most suitable for conversion into steel by the cementation process. The trade in iron centred naturally in Bristol, as did that in most other overseas raw materials that were required; materials came up the Severn to the point on the river nearest to the Black Country, to Bewdley, a town which owed its importance to the fact that it was close to the point where the river Stour flows into the Severn. Bewdley thus became the entrepôt for the trade up the Stour valley carried on by pack animals or by barge on the Stour and Salwarpe navigation already mentioned. Nash in his *Worcestershire* gives a vivid picture of the busy scene as late as 1795, of the arrival at Bewdley of these pack animals, or their congregation there, ready to be loaded for their destinations.

On the other side of the Black Country the same means of

transport of finished goods—by pack animals—was employed, to the rivers Dee and Mersey on the one hand or to the river Trent on the other.

What time the stage waggon came into the district to supplement and eventually to supplant the pack animal we have not been able to find, but there was a service to London in existence in the eighteenth century.

Within the period with which this volume is concerned, the coming of canal navigation to the district wrought great changes and conferred many benefits. The difficulties to be overcome were more than ordinary because of the differences in level, seeing that the area is more or less of a tableland as much as 400 ft. above sea level. The Staffordshire and Warwickshire Canal, authorized in 1766, with James Brindley as engineer, passed from the Severn to the summit at Compton, thence to the Trent and Mersey Canal at Great Haywood, a distance of 46½ miles. At its junction with the Severn, where the Stour debouches into it, Brindley created the new town of Stourport and thus gradually drew away all the heavy goods trade from Bewdley. In 1768 the Birmingham Canal was authorized; this passed through Bilston and joined the previously named canal at Autherley. In communication with these two canals, and in conjunction with trunk canals subsequently constructed, a complete network serving the whole of the Black Country was brought into being and it is a feature of the district that the sites of manufacturing works have been chosen close to a canal so that the works should possess a wharf whither coal and other bulky materials could be brought and whence goods could be dispatched. Indeed it may be said that no area in England is better served with canals than is this.

Intercourse of the inhabitants of the district with the outside world before the fourteenth century must have been practically non-existent, but whatever it was must have been accomplished both by men and women on horseback. The stage waggon and later the stage coach were long in penetrating into the district owing to the narrowness of the roads and to the state of their surface, fit only for horseback communication, which indeed had the advantage over any other means in point of speed. Celia Fiennes speaks feelingly of the state of the roads she experienced in this part of England about 1697, when describing an incident that took place near Wolverhampton:[1]

Here we had the inconvenience of meeting the Sherriffs of Stafford-shire Just going to provide for ye Reception of ye Judges and officers of ye Assizes. Whose Coaches and Retinue meeting our Company wch was encreased wth Cosen Fiennes' Coach and horsemen wch made us difficult to pass Each other in the hollow wayes and Lanes.

The private coach was followed at a long interval by the public stage coach. The first service of these to run regularly from Birmingham, or indeed any part of the district, to London was established in 1731. This was an event of such importance to the merchants and manufacturers, not to speak of the general public, that we must be pardoned for giving the notice[2] of the service in full. The notice is headed by a woodcut of a stage coach.

Birmingham

Stage-Coach

in Two *Days* and a half; begins *May* the 24th 1731.

Sets out from the *Swan-Inn* in Birmingham every *Monday* at six a Clock in the Morning through *Warwick Banbury* and *Alesbury* to the *Red Lion* Inn in *Aldersgate street*, London every *Wednesday* Morning: And returns from the *Red Lion Inn* every *Thursday* Morning at five a Clock the same Way to the *Swan-Inn* in *Birmingham* every *Saturday*, at 21 Shillings each Passenger, and 18 Shillings from *Warwick* who has liberty to carry 14 Pounds in Weight and all above to pay *One Penny a Pound.*

Perform'd (if God permit)

By Nicolas Rothwell.

[1] Fiennes, C., *Through England on a side saddle*, edited 1880, p. 194.
[2] Preserved in the Assay Office.

In 1742 the time of the journey to London was reduced from two and a half or three days to two days.

In 1783 there was a still further acceleration of the coach service from the Swan Inn, as witness this advertisement.[1]

> *THE* OLD ORIGINAL *LONDON* well known expeditious POST-COACH in sixteen hours every Evening at six oclock. Fare inside 1l 11s 6d

Up to that time the speed of travel had not been improved upon since Roman times. The astonishing acceleration that has taken place in the last 150 years owing to our command over the forces of nature is known to every one, but the cost of travel, it will be observed, has not been much reduced.

The question will now be asked why should Birmingham, of all the towns and villages which rose to importance in the South Staffordshire area, have outstripped the rest and have become the metropolis of the Midlands, rather than, say, West Bromwich, Wolverhampton or Stourbridge.[2]

Geologically speaking, Birmingham lies on the Keuper beds and the Bunter sandstone which not only furnish a good supply of well water, permitting of a large population, but at the elevation at which it stands, about 400 ft. above sea level, is an eminently healthy spot. Its situation on the banks of the brook Rea, near its confluence with the river Tame, was doubtless the fact that influenced its original settlement. The brook, primarily of value because of the water it furnished, became important later, as did other tributaries of the Tame, from the point of view of water power supply.

Birmingham, unlike the other towns mentioned, is not actually in the South Staffordshire coal field but is about 3 miles distant from its edge. This must have been an initial disadvantage because of the cost of transport to it of coal and iron, when we reflect that such cost by packhorse was of the order of

[1] *Aris's Birmingham Gazette*, July 14, 1783.

[2] In this connection the introductory chapter of *Industrial Development of Birmingham and the Black Country, 1860–1927*, by G. C. Allen, should be consulted.

2s. or more per ton per mile. This disadvantage was offset by the fact that for the transport of manufactured goods it was nearer than the others to Icknield Street, the old line of road passing through Northampton, Coventry, Lichfield, Stafford and Stoke, thereby giving access to London on the one hand and Chester on the other. As regards water carriage Birmingham was better situated than the towns mentioned, except Stourbridge, for access to the Severn by way of the Stour navigation for the importation of heavy raw materials or the dispatch of finished goods.

It was natural that the towns actually on the coal and iron fields should devote their energies to the primary stages in iron manufacture and that the skill of the workmen should be exercised and developed in that direction rather than in working up their product into finished goods. Such a division of labour must have come with time in any case, and it is to the credit of the Birmingham men that they should have realized that concentration on highly finished articles was what they were best fitted for. With the segregation of trades that we have mentioned, this division of labour would become stabilized.

Birmingham, although nearest of the towns mentioned to Icknield Street, was still about 15 miles distant at the nearest point and in the Middle Ages that was sufficiently far away for it to have remained, as it did, isolated to such an extent that it did not, like Coventry and the rest, attain to borough status. Hence Birmingham had no trade gilds or corporations, factors that were destined to be of importance in the capitalistic phase that had now been entered upon, for there were no such restrictions as elsewhere on enterprise.

In the corporate towns up to the seventeenth century apprenticeship was compulsory; a master or journeyman was restricted to a particular trade, these trades were elaborately specialized and rigidly demarcated, methods of production were prescribed, and wages and prices were regulated. In Birmingham on the other hand a man was never asked whether he had served an apprenticeship or not.[1]

[1] Cf. Tucker, *Essay on Trade*, 1753, p. 87.

The Civil Wars passed lightly over Birmingham, perhaps because it had not much to lose, and left little aftermath as compared with important places like Coventry which were severely mulcted in fines.

The author shares the opinion that economic factors are not the only ones to be taken into account, as is too much the fashion, in tracing the rise or fall of a community. The human element is important and in the case of Birmingham, so the author believes, all important. Freedom from restriction and authority has its good side and in her case it has resulted in the attraction to the town of men of advanced views—vigorous dissenters of all kinds, notably members of the Society of Friends (the Quakers) in whose hands, in the eighteenth century, so much of the iron industry was carried on. Development is cumulative in effect and the new industries that were brought in attracted not only clever artisans from a distance but also men of the mercantile class who were quick to see openings for their abilities. This continual infusion of new blood, reacting upon the existing inhabitants, led to a spread of technique, a quickening of ideas and a broadening of outlook that have been such distinguishing characteristics of the inhabitants of the town.

Most important of all, these factors led to the welcoming of men of invention, the source whence springs new industry. This fact is attested to by our patent law records. Taking as a convenient date for the purposes of this volume the passing of the Patent Law Amendment Act of 1852, we find that from 1617 to that date, Birmingham, of all the provincial towns, stood first as regards grants of patents. The nearest in point of numbers of patents granted was Manchester with only one-third of the number of Birmingham. How many of the patentees were indigenous and how many came from outside we cannot say, but we can say that the town, since it commenced its industrial career, has been a magnet for skill and brains and the attraction still continues. It is difficult to find a family there, even to-day, that can claim to have belonged to the city for more than a generation; most of its citizens are what in the North of England are called "offcomes".

It is in the conviction that Birmingham owes its position so largely to the men who have gone to work there that the author has been encouraged to put on record the life of one of such men as have been described—a man who did much for the town but who has never been accorded the meed of praise properly due to him—Matthew Boulton.

CHAPTER II

THE BOULTON FAMILY

Rise of the family and its coming to Birmingham—Birth and education of Boulton—Enters his father's business—His inlaid buckle—Marriage—Death of his father and of his wife—The buckle-chape bill—Second marriage—Comes into a fortune—Friends and contemporaries.

MATTHEW Boulton busied himself in many parts of the field of industry and in every one of them he left some mark. These occupations were successive in point of time and in nearly every one of them he achieved something of a lasting character. Each occupation as he took it up was pursued in parallel with those he had previously undertaken. We might liken Boulton to a chariot rider who starts with a single steed and at successive stages harnesses another one until finally he is driving a whole team abreast. It is these achievements that we shall describe in the following pages, without however losing sight of the man himself who is perhaps more interesting than his work.

Thus Boulton's life can be divided broadly into occupational periods. First in his young days he entered upon the manufacture of hardware; next in early life he took up enthusiastically artistic productions—Sheffield plate, ormolu, silver plate and pictures; then in middle life he threw himself into the development of Watt's steam engine, the most strenuous period of all and the most important; lastly in advanced age, with remarkable energy he brought about the improvement of coinage, undoubtedly the most congenial of all. Not till then, and hardly then, would he consent to say *Nunc dimittis*.

Before enlarging upon the theme, and in order not to depart from the accepted canons of biography, something must be said about the Boulton family, about Boulton's early years and about the manner of his setting forth into the world.

The rise of the Boulton family is typical of what took place commonly in the seventeenth century, viz. the arrival of a new class in the community, having its roots in the landed gentry and engaging capital and energy in manufacturing pursuits. John Bolton (for the name was then and subsequently written without the "u"), grandfather of Matthew, is the farthest back that we have been able to trace. This John is said to have been a descendant, perhaps a grandson, of Robert Bolton (*b.* 1572), the Puritan divine, rector of Broughton, co. Northants, from 1609 until his death. The parish register records his burial thus: "egregius ille concinator idemque Rector de Broughton sepulchro compositus Decemb. 19 1631". In the parish church there is a remarkable monument to him representing him in the pulpit.[1] No confirmation of this tradition is forthcoming, however. John Bolton married Elizabeth, daughter and eventual heiress of Matthew Dyott of Stichbrooke, co. Stafford, and of Mary his wife, a Babington of Curborough, a family that has left its mark on English history. It is quite possible that this marriage with an heiress may have established the fortunes of the branch of the Bolton family and enabled John to settle, as he did, in the rising town of Birmingham, there to set up in business early in the eighteenth century.

John's son, the elder Matthew, was in business, like so many others, as a toy maker. This occupation was not that of a maker of children's toys as anyone at the present day would conclude, but a maker of the hardware that led Edmund Burke to call Birmingham "the toy shop of Europe". The word toy in this sense is used with a somewhat wide connotation, so we cannot do better than quote a contemporary definition:[2]

Toy Makers.

An infinite Variety of Articles that come under this Denomination are made here; and it would be endless to attempt to give a list of the whole, but for the information of Strangers we shall here observe that these Artists are divided into several Branches as the Gold and Silver Toy Makers, who make Trinkets, Seals, Tweezer and Tooth Pick cases, Smelling Bottles, Snuff Boxes, and Fillegree Work, such

[1] Information by the kindness of the present rector, the Rev. C. Podmore, M.A.

[2] *Sketchley's Birmingham Directory*, 1767, p. 56.

as Toilets, Tea Chests, Inkstands &c. &c. The Tortoiseshell Toy maker, makes a beautiful variety of the above and other Articles; as does also the Steel; who make Cork Screws, Buckles, Draw and other Boxes; Snuffers, Watch Chains, Stay Hooks, Sugar Knippers &c. and almost all these are likewise made in various Metals.

It is not to be imagined that Matthew Boulton made more than a very few of the bewildering number of articles mentioned above. Boulton's branch of the trade was that of the Steel Toy maker. It was usual for a maker to confine himself to a few articles within his own branch and there is some likelihood that Boulton made only buckles. Probably too his labour force was small, for that was the usual way of working then. The maker did not market his product; he worked to the order of the merchant or middleman who had his correspondents and supplied customers in this country and abroad. Boulton's place of business was in Snow Hill, then quite on the outskirts of Birmingham on the road to Wolverhampton, described as "a very rural neighbourhood, only sparsely built upon, in close proximity to the open country". Some stress has been laid upon the fact that his workshop was there rather than with others of the same trade who were in the centre of the town, and therefore that he must have been in a small way of business. We have already assumed that the latter was the case but as far as premises were concerned it seems likely that, as a newcomer, he would have to take what he could get and Snow Hill was the natural outlet.

Of the elder Matthew, beyond that he married Christiana, daughter of a Mr Piers of Chester, we know very little but we have one link with him in the shape of a small mail or hair trunk, in which he kept his papers, now in the possession of his great-great-grandson, Major-General Sir Frederick Robb.

There were several children of the marriage, but the only one who made any mark in the world was Matthew, the subject of the present biography. Family connections are not readily grasped without the aid of a family tree and such a one brought down to the present day is therefore appended (see p. 27).

Matthew Boulton was born on September 3rd, 1728 (September 14th, New Style)[1] in the dwelling house behind the workshop in Snow Hill abutting on Slaney Street. The premises can still be seen. He was baptized on September 18th, in the parish church of St Philip, now the cathedral church of the diocese of Birmingham.

It was a source of satisfaction to his practical mind, when he was grown up, that he had been born in that year, for the figures of the year are the same as those of the number of cubic inches in a cubic foot.[2] He liked to refer to this fact among his intimates and we find him mentioning it in his letters (see p. 119).

We know very little of Boulton's boyhood and early years. He went to school at the academy of the Rev. John Hausted, who was chaplain of St John's Chapel, Deritend, from 1715 till his death in 1755. In his obituary notice we are told that he was "master of a Private School", and "remarkable for his great Abilities as a Divine, and for his Learning and unwearied Diligence in the Instruction of the youth committed to his Care". Making all allowance for the desire to speak well of the departed, we can conclude that he was an able schoolmaster and that into one pupil at least he had instilled some of his learning, and, what is more, a desire to go on learning. Boulton shows in his correspondence some acquaintance with the classics and a knowledge of English literature, yet we cannot doubt that this was due in no small measure to his own efforts in pursuing his education after he left school.

It may be asked very pertinently why the boy was sent to

[1] Persons who were born in England and Scotland before the introduction of the Gregorian calendar in 1752 whereby the days September 3rd to 13th were omitted, and survived the change, kept their birthdays 11 days later than the nominal date of birth. Thus Boulton after 1752 kept his birthday on September 14th (see p. 119).

[2] Whether it was pride in this fact or because the year was so important for the fortunes of the family, Boulton's grandson, Matthew Piers Watt Boulton, who in the later prosperous days of the business was seated at Tew Park, Oxon, had all the cottages in the village of Great Tew reconditioned by an architect and a date stone "1728" inserted in each. This village of thatched cottages is to-day one of the most picturesque in the kingdom.

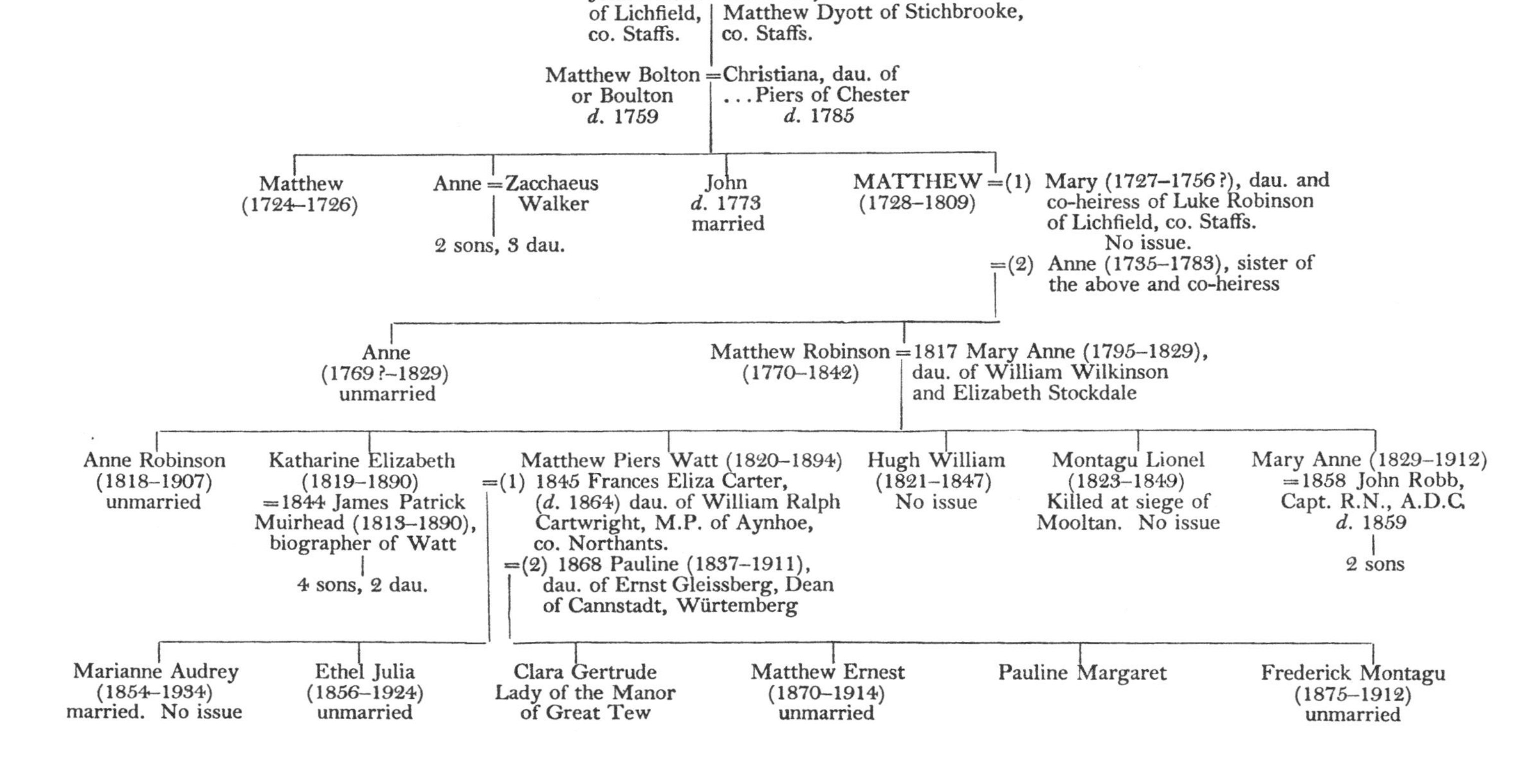
BOULTON PEDIGREE
John Bolton = Elizabeth, dau. and co-heiress of
of Lichfield, co. Staffs.
Matthew Dyott of Stichbrooke, co. Staffs.
Matthew Bolton = Christiana, dau. of
or Boulton
d. 1759
. . . Piers of Chester
d. 1785
Matthew
(1724–1726)
Anne = Zacchaeus Walker
2 sons, 3 dau.
John
d. 1773
married
MATTHEW = (1) Mary (1727–1756?), dau. and co-heiress of Luke Robinson of Lichfield, co. Staffs.
(1728–1809)
No issue.
=(2) Anne (1735–1783), sister of the above and co-heiress
Anne
(1769?–1829)
unmarried
Matthew Robinson = 1817 Mary Anne (1795–1829),
(1770–1842)
dau. of William Wilkinson and Elizabeth Stockdale
Anne Robinson
(1818–1907)
unmarried
Katharine Elizabeth
(1819–1890)
=1844 James Patrick Muirhead (1813–1890), biographer of Watt
4 sons, 2 dau.
Matthew Piers Watt (1820–1894)
=(1) 1845 Frances Eliza Carter, (d. 1864) dau. of William Ralph Cartwright, M.P. of Aynhoe, co. Northants.
=(2) 1868 Pauline (1837–1911), dau. of Ernst Gleissberg, Dean of Cannstadt, Würtemberg
Hugh William
(1821–1847)
No issue
Montagu Lionel
(1823–1849)
Killed at siege of Mooltan. No issue
Mary Anne (1829–1912)
=1858 John Robb, Capt. R.N., A.D.C.
d. 1859
2 sons
Marianne Audrey
(1854–1934)
married. No issue
Ethel Julia
(1856–1924)
unmarried
Clara Gertrude
Lady of the Manor of Great Tew
Matthew Ernest
(1870–1914)
unmarried
Pauline Margaret
Frederick Montagu
(1875–1912)
unmarried

Hausted's Academy rather than to King Edward VI's Grammar School, a foundation that had been in existence since 1552; the reply is that the school was then in a semi-moribund state. "Edward Mainwaring, of Preston, Clerk" had been appointed headmaster in 1726 and held the post for twenty years; during that time here as elsewhere in the country the decay in educational foundations that we find so commonly in the eighteenth century went on almost unchecked. Indeed it would appear that in Boulton's boyhood, Mainwaring had no scholars at all; the Academy therefore must have been a case of Hobson's choice.

It is idle perhaps to speculate what would have been Boulton's career if his life had been cast in another century than it was. If born in the Elizabethan period, for example, we can imagine that he would have used his family connections to get to Court, to intrigue there, to pull off a profitable monopoly or else to equip a privateer "to singe the King of Spain's beard", with a thousand per cent. profit to himself and to the bold sea-dogs who accompanied him. But the age in which he was born was utilitarian and high emprise was to be found readily only in mercantile pursuits.

Boulton left school presumably when he was fourteen years old for that is the age when it was usual then to do so, either to become apprenticed to a trade or to enter into business. It is hardly necessary to say that young Boulton's predilections were for the latter career and his father's business being invitingly open, it was the natural and almost inevitable course for him to enter it. The youth seems to have thrown himself into it with great energy for he was of an enterprising disposition.

The paternal business was, as we have said, that of toy making, and the principal product was buckles, both for shoes and knee-breeches. These were exported to every country of Europe where fashion reigned, particularly France, whence they were reimported into this country as the *dernier cri*. They had been the vogue for more than a quarter of a century and the Birmingham makers had not only kept pace with the vagaries of fashion in the size, material and ornamentation of the buckles but were themselves in the forefront in bringing out novelties. Young

Boulton was in no way behindhand and when only seventeen years of age he is said to have produced an enamelled or inlaid buckle which helped greatly to enlarge the output of his father's business; what exactly the invention consisted in we have not been able to find out, but we hazard a suggestion later when we learn more about buckles.

When he came of age his father took him into the business as a partner and from that time onward entrusted his son with the entire management. Their premises were, as we have said, on Snow Hill and there is a tradition that they were originally on a cramped site on the side now covered by the Great Western Railway Station and that the removal to the other side of the road was occasioned by the rapid expansion of the business due to the energy of the junior partner. There is no evidence, however, that the Boultons had manufacturing premises anywhere else than on the spot abutting on Slaney Street that we mentioned first. Their residence was in New Hall Walk (now Colmore Row), at that time the approach to New Hall, a stately but decayed old manorial residence to which access was gained from New Hall Walk by a fine avenue of trees represented to-day by Newhall Street.

The young man was of an enquiring turn of mind; as illustrating the latter trait we may mention that a paper happens to have been preserved in which he had garnered information about poultry keeping[1], whether with the idea of starting an intensive system of egg-production or not we do not gather. About the same time he was busy making thermometers for himself and friends.

The young man had good looks, a frank, engaging manner and was a good mixer; he appears to have got on well with his fellow manufacturers in Birmingham and district, of whom there were more than a hundred engaged in the same line of business as himself. With charm of manner and sociability of disposition it was inevitable that the eligible bachelor should be much in

[1] Boulton had read Reaumur, *The art of hatching and bringing up domestic fowls of all kinds at any time of the year, either by means of the heat of hotbeds or that of common fire*, Eng. trans. 1750.

request, particularly in his own circle, and it is not surprising therefore that he became engaged to be married. We shall accuse Boulton of no greater worldliness than that of the Quaker who advised his son "not to marry for money but to go where money is" when we say that the young man's choice fell upon Mary, the elder daughter of Luke Robinson, the "opulent mercer" of Lichfield. Mary's mother Dorothy was the daughter and co-heiress of John Babington of Curborough and of Packington, co. Stafford. It will be remembered that Boulton's great-grandmother was a Babington, so that Matthew and Mary were distant cousins. In a period when family connections were more thought of, if no less valuable than at the present day, we must admit that the match was an excellent one. Curiously enough, we have failed to find out when and where the marriage took place, but we believe it to have been about 1756 when Boulton was twenty-eight years of age.

They had not been married many years when a great misfortune befell the husband through the sudden death of his wife; we have been unable to lift the veil enshrouding the sad event, but circumstances point to its having occurred in 1759 or 1760. There were no children of the marriage, or at any rate none that survived. The loss to Boulton was a poignant one if we are to judge by a scrap of paper which has been preserved and on which he has written: "Upon seeing the corps of my dear Wife Mary many excellent Qualitys of Hers arose to my mind which I could not then forbere acknowledging extempory with my pen & depositing it in her Coffin." To this is attached a copy of his prayer to the Almighty for mercy.

He received another blow about the same time through the death, in 1759, of his father, the elder Matthew. He is said to have left his property to Matthew junior, but probably he left the business only, for he would surely have made some provision for his wife Elizabeth—she survived till 1785—and probably for the other children of the marriage. It would have been interesting to have learnt what were the actual provisions, but no document survives and a search for his will has proved fruitless.

The death of the elder Boulton naturally brought the younger one into greater prominence than before. His address and energy, coupled with a certain diplomacy that he exhibited from the outset of his career, seems to have brought him to the front among his fellow manufacturers not only in Birmingham but also in its neighbourhood, for we find him prominent among the buckle makers in petitioning the House of Commons in 1760 for leave to bring in a bill to prohibit the exportation of buckle-chapes.[1] As the term must be unfamiliar to most of my readers, let it be said that the chape is the tongue or part of the buckle by which it is fastened to the strap or ribbon. Chape making and buckle making were independent trades and it is interesting to see how far and how quickly division of labour had gone by this time.

The interesting part, to us, of the proceedings in the House is that Boulton appeared on April 22nd, 1760, before a Committee of the whole House as a witness for the petition, and as a representative not only of the buckle makers of his own town but also of those of Warwick and Wolverhampton. He was then only thirty-two years of age; it may even have been his first visit to London, but whether that was so or not it must have resulted in broadening his horizon in many respects.

The report[2] of Boulton's examination by the Committee is so interesting for its revelations as to the trade generally and as a picture of his own activities at Snow Hill in particular that we are tempted to give it in full:

Mr *Matthew Bolton*, a Toymaker of *Birmingham* being examined, said, That upwards of Eight thousand people are employed in Buckle-making in the Counties of *Warwick* and *Stafford*, the Returns of which are £300,000 *per Annum* exclusive of the Hands employed in making Tools, and preparing Materials: That Buckles are composed of Copper, Brass, Iron, Tin, Spelter; and that great Quantities of Buckles are set with Glass, in imitation of Jewelry: That on an Average, Chapes, are One Third of the Value of the ordinary Sort of Buckles; but there are Buckles made, that the Chapes are not One Tenth of the Value:

[1] *House of Comm. Journ.* vol. 28, 1757–61, pp. 785–901.

[2] *Loc. cit.* p. 882.

That the Home Consumption of Buckles bears no Proportion to the Exportation, as the greatest Part of the £300,000 is exported out of the Kingdom:

That the Witness believes there are about 2,500 Iron Chape-makers in the counties of *Warwick* and *Stafford.*

That the price of Chapes has been raised £15 *per Cent* within these Eighteen Months, owing to the great Demand for Exportation.

That if the Exportation of Chapes was prohibited, and they could be had at a reasonable Price, it would prevent Foreigners getting them, and by that means keep the Buckle Trade at home, and that he does not apprehend it will diminish the Number of Chape-makers:

That the Chape-forgers can get 20*s.* *per* Week, and the Filers from 9*s.* to 15*s.* *per* Week: That Chape-makers make much better Wages than other Manufacturers in the hard Ware, and may be kept constantly employed, but they will work but Four or Five Days in a Week:

That the Chape-makers have entered into Combinations, and have declared they should raise their Price 3*s.* in the Pound; and that Chapes, that were formerly sold at 18*d.* *per* Dozen, are now 1*s.* 9*d.* and not so good: and that this Rise is owing to the foreign Demand:

That he thinks Chapes cannot be made abroad, as they have not many Materials sufficient for that Purpose: That the Chape-makers give the preference to *English* Iron and will not use foreign Iron; but that Foreigners might endeavour to get *English* Iron, if they knew the difference between that and foreign Iron:

That there are Secrets in the Buckle Trade, which Foreigners are Strangers to, which induces him to think that Buckles cannot be made abroad so well as in *England.*

That if the Export of Chapes was prohibited, the Exportation of Buckles would increase, and as many Hands be employed in Chape-making as at present.

And, being asked the Question, said, That the Materials and Profits of the Master Buckle-maker is almost Half; for if a Master employs Forty Men, and gets £40 *per* Week, he pays £20 to his Journeymen, and that Journeymen's Wages is as £20 to £15.

Other persons were examined, including John Baskerville the printer.

Another witness said that "the Buckle is by far the most considerable Article in the Toy Trade".

Mr *Bolton* being further examined, said,

That the Orders for Chapes have greatly increased of late Years; and that the greatest part of the Exportation has been in the common Sort of Buckles; and he produced to your Committee Specimens of Several Sorts of Buckles:—

That they cannot make Chapes in *Spain* and *Portugal* so good and cheap as in *England*, as they have no slitting or rolling Mills; and if they were to import Iron there, it would be of too high a Price for the Manufacture; and they have neither Workmen nor Coals: That the *Germans* can make Chapes; but they take a great quantity of us, as their Iron is not so fit for Chapes; and if they could make Chapes sufficient, they would not buy of us; and if we were deprived of our Chape-makers we should be a long time before others could be instructed; and by that means the Trade would be lost. There must be some invincible Impediment to making Chapes in *Germany* because if they could make them they would make them £100 *per Cent* cheaper than we.

This account of buckle and chape making, the number of persons employed in this industry, the conditions of labour, their wages and the value of the product, are of great interest to the historian of economics. We conclude that the buckles "set in Glass in imitation of Jewelry" (i.e. enamel) are those which Boulton himself invented.

The buckle-chape makers responded with a counter petition against the Bill. The procedure that is familiar was followed—their petition was submitted to the same Committee and considered; after deferring their recommendations several times, the Committee reported against the Bill and on May 1st it was dropped. The constitution of the House, unreformed as it then was and without any representatives of industry whatever, eliminated trade interests and made this fair decision easy. It is difficult to imagine a more impudent attempt on the part of one branch of a trade to try and restrict another branch from doing exactly what the first was doing, with the obvious intent thereby to cream off the lion's share of the profit. It is to be hoped that Boulton lived to regret the illiberal part he had played in promoting the Bill.

Young Boulton appears to have taken the loss of his father philosophically, but not so that of his wife. We can imagine his still keeping up his visits to Lichfield to seek the sympathy of his mother-in-law of whom, judging by affectionate letters to her that have been preserved, he was fond. It is easy to understand that he would be attracted by the younger Robinson daughter, Anne, seven years his junior; we may even hazard the conjecture that it had been from the first a case of Leah and Rachel. He would be deterred, however, from thinking of marriage with Anne by the knowledge that by ecclesiastical law, very different from the customs of patriarchal times, marriage with a deceased wife's sister was forbidden.

That Boulton was turning over the matter in his mind seriously is proved by a curious piece of circumstantial evidence which incidentally fixes the approximate date of the death of his first wife. The evidence is this: there has been preserved an account to Boulton from J. Whiston and B. White, publishers and booksellers "at Boyle's Head in Fleet Street" dated 1760, Apr. 22, for supplying 180 copies of "Fry on Marriage"[1] at a cost of £9. It is difficult to explain why Boulton ordered so many copies. The only explanation that suggests itself is that he wanted to distribute copies to everyone to whom his marriage was a matter of concern.

The author *inter alia* arrives at the conclusion (on p. 80) "that marriage with a deceased wife's sister is fit and convenient being opposed neither to law nor morals". The inference is that this book influenced Boulton in arriving at the same conclusion as the author and in justifying him in making advances to the younger sister. That he was ardently and passionately in love with "Miss Nancy" as he called her we are in no doubt, for his love letters, which have been, probably accidentally, preserved, are an indication of the fact.

The tradition is that the Robinson family were much opposed to the match; and well they might be, not for the reason always

[1] Fry, John, *The case of marriage between near kindred particularly considered with respect to the doctrine of scripture, the law of nature and the laws of England*, 1756, Price 2*s*., published by J. Whiston and B. White. It will be observed that the businesslike Boulton had secured a discount of 50 per cent. on the published price.

given that there was a difference in station between them, for, as we have seen, there was none, but because marriage with a deceased wife's sister was contrary to the Table of Kindred and Affinity in our Prayer Book, and was considered morally wrong in the eyes of ninety-nine persons out of a hundred at that date.

Existing letters show that Anne's brother Luke and perhaps the father were opposed to the match, but it does not appear that the mother was, and she, being the heiress and fond of Boulton, was an important person. We do not know when the marriage took place—without doubt it was kept secret—but we infer that it was prior to 1767. We base this conclusion on a letter to Boulton written (1767, Nov. 4) by Mr H. Hassall, attorney-at-law of Lichfield in which the latter mentions that he finds the "Reversion in fee (i.e. of the Packington estate) is vested in the late and present Mrs Boulton"; while Boulton, writing (Nov. 24) to Hassall, refers to legacies that "centre in me and my present wife". This is fairly conclusive but there is further circumstantial evidence to corroborate it.

It would appear that the relationship of Boulton's wives was known to his friends, for Samuel Garbett, writing to him from Carron (1772, Jan. 5), says:

I congratulate you on your discovery at Packington (his wife's estate there), I am sure you have need of all your good sense to be on your guard in that quarter of the world, and particularly in case of Mrs Boulton's death. I have often feared for your little folks (i.e. Nancy and Matt) that you have not been sufficiently decisive on that point, and therefore never think of it but with a sensation you would wish for a hearty friend, which I know you will accept as an apology for my saying anything about it.

Further evidence that the relationship of the wives was known is afforded by correspondence with Richard Lovell Edgeworth (1744–1817) whom we shall meet again as a member of the Lunar Society; being desirous in 1780 of marrying his deceased wife's sister, he wrote (Oct. 2) to Boulton asking for the latter's advice on the matter and Boulton replied (1780, Nov. 20):

The only advice I can give you upon the subject of your letter is to read a small book upon Marriages between near kindred by Fry.

When you have taken your resolution I advise you to say nothing of your intentions but go quickly and snugly to Scotland or some obscure corner in London, suppose Wapping, and there take Lodgings to make yourself a parishioner. When the month is expired and the Law fulfilled, Live and be happy. The propriety of such a marriage is too obvious to men who think for themselves to need my comments....I recommend Silence, Secrecy & Scotland.

It only remains to say that Edgeworth followed Boulton's advice and married his wife's sister at St Andrew, Holborn, on Christmas Day, 1780.

It would be hypocritical to pretend that Boulton was uninfluenced in his second marriage by the fact that the sisters were co-heiresses under their father's will. Since by the law of that time the property of a woman became that of the man she married, Boulton may have thought that he might succeed in getting control of both fortunes. This, it appears, is what did happen, but how he did so in the case of the second marriage it is difficult to say, for there must have been legal difficulties in the way. Until we know more of the circumstances we can only hazard the suggestion that Anne Robinson made over her fortune to Boulton by deed of gift. The fortune of each daughter was about £14,000 in landed estate. This fortune, or at any rate some of it, must be that referred to by Benjamin Franklin who, writing to Boulton in 1765 (May 22), says: "Mr Baskerville informs me that you have lately had considerable addition to your fortune on which I sincerely congratulate you." How this came about at this particular juncture we do not know. Possibly the fact that Boulton's brother-in-law Luke Robinson died in 1764 may have some bearing on the question, on the supposition that he had some life interest in the estate.

By the time the fortune did come to him, as we shall find in the succeeding chapter, he was deeply immersed in business. He might then have relinquished business altogether, retired to his estates and become the country gentleman. He said so himself towards the end of his life in reply to an insolent letter of Philip Thicknesse (1719–92), traveller, author, and somewhat of an upstart, who tried to insult Boulton by addressing him as

"Tradesman of Birmingham". The latter's dignified reply was as follows:

> Early in life Fortune gave me the option of assuming the character of an idle man commonly called a Gent^{n}, but I rather chose to be of the class wch Le Baron *Montesque* describes as the constant contributors to the purse of the commonwealth rather than of another class which he says are always taking out of it without contributing anything towards it.

But the life of a country gentleman had not the least attraction for Boulton, indeed the affairs of active business were to him as the breath of his nostrils and the only use that he could envisage for the fortune that he had received on marriage was to turn it into the wherewithal to prosecute the ambitious schemes that he had already entered upon. It was characteristic of him that he always wanted to be in the forefront, a leader of men and a captain of industry. Coupled with this was the desire that every product to which he set his hand should be superior in design, material and workmanship, if possible, to that which anyone else produced. It was unfortunately true that the goods produced in Birmingham at that period had a bad reputation because of their showiness, and poor quality; so much so that the name of the town in the popular pronunciation—Brummagem—was applied to its goods as a synonym for gaudy and worthless. This was a serious handicap to Boulton but he set himself resolutely to remove the stigma in so far as he was concerned.

By the time the fortune came to him and he was able to indulge the social and hospitable instincts that were so strong in him Boulton's circle of business men and manufacturers had widened considerably. He had too, as we know, through his family connections a circle of friends. As many of them greatly influenced him, it is desirable to say something of the more prominent of them.

Head and shoulders above the rest was Dr Erasmus Darwin (1731–1802) of Lichfield, the generous, big-souled and big-bodied free-thinking and radical physician. He was medical attendant to half the families in Warwickshire, including the family of Luke Robinson; no doubt it was through him that the introduction was effected; not that he was Boulton's physician, however, for Dr William Small (1734–75) appears to have acted in that capacity.

Dr Small was a man who had seen the world, for he had been Professor of Mathematics and Natural Philosophy in the College of Williamsburg, Virginia; the climate had not suited him, however, and he returned to this country in 1765 with an introduction to Boulton from Benjamin Franklin describing Small as "an ingenious philosopher and a most worthy honest man". The outcome of this was that Small settled in Birmingham and became a close friend and counsellor on whose sound judgment Boulton placed great reliance.

How Franklin (1706–90) came to know Boulton is not clear but it happened during the former's second stay in England, 1757–62, probably owing to a chance visit to Birmingham. They continued to correspond on matters of natural science and during his third stay, 1764–75, several letters passed between them relative to the construction of an atmospheric engine that Boulton intended to erect at Soho; of this engine we shall have something to say later.

Among Boulton's more strictly business friends we must mention John Roebuck (1718–94), M.D. of Edinburgh and of Leiden, one of our great captains of industry. He was one of our early industrial chemists, for he found sulphuric acid a laboratory product made in glass retorts and he advanced its manufacture to the industrial scale with great changes in technique in leaden chambers. He established a works for its production in 1749 at Prestonpans, Midlothian. Probably one reason for choosing a site in Scotland for his works was because he had a Scotch, but not an English, patent for the process. In conjunction with his partner Samuel Garbett he improved methods of refining gold and silver. Together with other partners they established the first iron works in Scotland to use local ores, at Carron, Stirlingshire, in 1760. Not content with that, he leased coal mines and established salt works at Borrowstounness (Bo'ness), Linlithgowshire, in which he desired Boulton to be associated as a partner. In these later projects

WILLIAM SMALL, M.D.
(1734–1775)
Professor of Natural Philosophy

JAMES WATT, LL.D., F.R.S.
(1736–1819)
Civil Engineer

JAMES KEIR, F.R.S.
(1735–1820)
Soldier and Chemist

JOHN WILKINSON
(1728–1808)
Ironmaster

PLATE I. SOME OF BOULTON'S FRIENDS AND ASSOCIATES

unfortunately he was unsuccessful and this led to his bankruptcy; this in turn had a material influence on the course of development of the steam engine and on Boulton's connection with it, as will appear later.

Samuel Garbett (1715–1803), whom we have just mentioned, was perhaps the most public-spirited man in Birmingham and acted frequently as a link between the manufacturers there and government departments in London. He was a partner of Roebuck and carried on their laboratories till the end of the century. In 1770 he was carrying on business as a merchant at 11 Newhall Street. Unfortunately he took into partnership his son-in-law Charles Gascoigne, who defrauded Garbett shamefully and brought him to bankruptcy; this cast a shadow over his career which lasted the rest of his life, although in his old age he recovered financially and paid his creditors every penny.

A man of like enterprise and animated by similar ideals to Boulton was Josiah Wedgwood (1730–95), the potter; he had established himself in business at Burslem, co. Staffordshire, in 1759 and greatly improved the ordinary delf ware. In 1769 he perfected the cream or Queen's ware—he was appointed Queen's Potter in 1762—with which his name is associated. In 1769 he established the famous pottery works at Etruria. He and Boulton were associated in production of fictile art objects and later when larger issues of trade and taxation arose they were associated in public life.

The portraits of some of the friends and associates of Boulton, mentioned above or to appear on the scene later, are shown on Plate I. The portrait of Small is from a pencil drawing, artist unknown, found among the Boulton papers in the Assay Office, Birmingham. It has not been published hitherto and is believed to be the only portrait extant. The portrait of Watt is from a replica of the well-known oil painting by Sir William Beechey, R.A., 1801. The original is in the possession of the Watt family. The portrait of Keir is from an engraving by W. H. Worthington of a pastel by L. de Longastre, *c.* 1805, believed to be in the possession of Keir's descendants. The portrait of Wilkinson is from the oil painting, artist unknown, in the possession of the Borough Council of Wolverhampton and hung in the Town Hall there.

We have dwelt rather too long on Boulton's family and social life, and have not kept pace with his industrial career in which he had been making great strides. We must now revert to this and tell how his ever-expanding business impelled him to remove to a new place of manufacture and to embark on an important partnership.

CHAPTER III

SOHO MANUFACTORY

Origin of the word Soho—Partnership with Fothergill—Toys and Buttons—Enticed to go abroad—Sheffield plate—Ormolu Clocks—Royal patronage—Silver plate—Founds Birmingham Assay Office—Visitors to the Manufactory.

THE ambitious scheme mentioned in the last chapter that Boulton now envisaged was to combine the activities and secure the profits of both manufacturer and merchant, a form of vertical integration of industry that most persons believe to be of modern origin. To carry out this scheme, his idea was to establish a manufactory of ample floor space supplied with water power where workmen of the different hardware and toy trades should be brought together under one business and technical management: overhead expenses could be reduced; by subdivision of labour and co-operation between different branches of the trade novel articles could be produced; by warehousing the products and merchanting them himself at home and abroad he could secure the merchant's profit and could command an ever-widening market.

Already in the Birmingham trade there was minute subdivision of labour; each branch was carried on by a small master with possibly half a dozen workmen on piece work and without the aid of any except muscular power. The work was done to order of the merchants who secured and distributed the orders, in fact organized the selling and export of the products. The establishment of a manufactory such as described was a step of profound import in industrial development. It had within it, as a moment's thought will reveal, the germ of what we know to-day as mass production. It must not be inferred that the idea was entirely novel for something closely analogous had been practised in the silk industry at Derby since about 1715, if not in other places.

Early in 1761 Boulton looked about him for a convenient site and found what he wanted on Handsworth Heath beside the road to Wolverhampton at a distance of $1\frac{3}{4}$ miles from Snow Hill. The site was outside the parish of Birmingham and not even in the same county—Warwickshire—but just over the border in Staffordshire. Until 1756 the land had been nothing but a heath covered with ling and gorse such as we described in the first chapter. The only building on it was the cottage of the warrener in charge situated on an eminence afterwards occupied by the dwelling house occupied by Boulton. The great advantage of the site was that it was traversed by the Hockley Brook, a tributary of the Tame. Boulton was not the first to recognize the advantages of the site, for Mr Edward Rushton had already taken a lease of it from John Wyrley, Lord of the Manor of Handsworth, had dammed the brook to form a pool, had built a rolling mill for metals and had erected a dwelling-house. However, we cannot do better than let Boulton himself tell the story of his conception, how he realized it and how he entered into partnership with John Fothergill. We abstract it from a memorandum entitled "Case between B. & F." and drawn up, to judge by internal evidence, in 1781. The memorandum is as follows:

About Jany 61 B had conceived a plan for ye manufactoring of various articles in ye Birm^{gm} Hardware & toy Trade in wch a water mill was essentially necessary & that as near to the town of Birm^{gm} as possable. He according apply^{d} to Mr. Edwd Ruston who had at in y^{e} y^{r} 56 taken a Lease (for 1 Hun^{d} yrs) from J. W. Esqr of certain Lands and priviledges situated in y^{e} parish of Hansworth ab^{t} $1\frac{1}{2}$ miles from Birm^{gm} & in that & ye following yr did make a Canal full half a Mile in Length to convey a little Brook to a place for ye working of a Mill wch he accordingly erected with a dwelling House for himself at ab^{t} 150 y^{ds} distance from ye mill wch with other lasting improvements made upon ye s^{d} premises it appeared by his accounts that he had expended the sume of £1000. Now as this situation appear^{d} more eligable than any other for Mr B['s] intended plan he therefore agreed wth Ruston & did purchase his Lease of him wth all his improvements for ye sum of 1000 but as ye Mill was ye only object of Trade & M B being desireous of keeping the Acc^{ts} of Trade as distinct & clear as possable he therefore opend two accts in his Books Soho Mill & Soho House wch two acc^{ts} then became D^{r} to his cash for 1000 p^{d} Edwd Ruston for his Lease viz

Fig. 1. Part plan of the township of Handsworth, 1794.
Courtesy of the Public Libraries Committee, Birmingham.

To Walsall
SOHO HOUSE
SOHO
ANUFACTORY
I R M I N G H A M
To Birmingham
NS
20
30
1000
2000
T

Mill &c at Soho for 700 & House &c at Soho 300 wch was as near each of their original expences as could be guesed. After wch M B was carefull to keep a distinct acct of ye Mony Layd out on either of these accts & did accordingly expend a larg sum same yr in finishing ye House within (wch before was almost bare walls) & in makeing a new Kitchin Garden building planting above 2000 Firs & a great variety of Shrubs wch in ye whole amtd to abt 500 add[it]ional expence to ye House but ye Mill being M Bs great object he labourd assiduously to perfect & perticularise ye several parts of so larg a machine. Abt July ye sd summer 61 he had procured ye necessary workmen Timber Brick &c & did according[ly] then begin to Build some dwellings for Workmen a Warehous wth several shops & as the old construction of ye Mill was ill favd for M Bs new plan he thought it most convenient and prudent to pull down ye old Mill wch he accordingly did & rebuilt it but as these Building were not finished that year (viz 1761) & has Mr B had been at ye expence of Building a Brick kiln & opening a mine of Clay & thought it proper to agree wth his Brickmaker to make more Bricks in the Spring to compleat ye Buildings then begun & to sell the overplus—It happe[ne]d that M B had a call to London in Janry 62 during wch time some difference arrose between F[othergill] & his old Master Mr Duncumb a merchant in Birmgm & in consequence an immediate seperation ensued. 2 days after F set out for London to propose a partnership with me wch I then declined but upon repeated application of himself & Friends B Consented to the proposed union M B & F then dayley conferd together upon the terms of their Partnership F told B he would be content wth $\frac{1}{3}$ Share of the profits & advance equal sums in consideration of M Bs superior Skill & of the greatest Burthen of ye Business naturely falling upon him.

The memorandum sets out clearly the developments that Mr Rushton had made in the estate before Boulton came on the scene and these will be still more readily grasped by reference to the map (see Fig. 1). This is a portion of "A plan of the Township of Handsworth in the County of Stafford, Drawn in the year 1794 by Saml. Betham", i.e. when Handsworth Heath was being enclosed. The relationship of the estate to the main road from Birmingham to Wolverhampton, close to the junction of the road to Walsall, the "canal" or mill lade that Rushton cut and the bay or mill-dam that he formed, together with the old course of Hockley Brook at this point are indicated.

The name given to the estate was the foreign-sounding one of Soho, but when it was so christened and what is the origin of the name are obscure although there are several traditions current as to these points. The first mention of the name that we have found is in a deed of 1761 (not executed, however) which recites the lease of property by John Wyrley, Lord of the Manor, in 1757 to Edward Ruston and John Eaves comprising that "warren in Handsworth aforesaid called Soho warren" and "crabtree brake warren". This does not, however, explain the origin of the name although it discredits one tradition, i.e. that Boulton himself conferred the name on the estate. Another tradition that it took its name from the sign of an inn on the Heath is vouched for in the following note[1]:

> ...A brick house with its front towards Birmingham, stood in the Park, about 20 yards past Hamstead Road, and about 15 yards from the roadside. It was there till the Park was cut into for building purposes. I was told 50 years ago...that this house, before the Park was inclosed was a public-house and had for its sign a representation of a hunt, there was a hare, hounds, horses and men, and the huntsman was represented with the word "Soho" proceeding from his mouth....

This is the tradition that is credited by Mr R. K. Dent, the Birmingham historian.[2]

It will be observed that the tradition carries us no further back than 1817. However, if the map of 1794 be examined, it will be seen that there *was* a house in exactly the position stated, at the junction of the road to Walsall (subsequently known as Hamstead Road) and in quite the best position for a public house; the house is enclosed within its own fence. The tradition is that the inn sign represented a huntsman urging on his dogs and that the word "So-Ho", "So-Ho" was a hunting cry. One cannot help recalling that a closely similar tradition exists as to the origin of the name Soho in London; hence the suspicion arises in the mind that the tradition may have been transplanted

[1] *Birmingham Journal*, Feb. 23, 1867, "Notes and Queries".

[2] *The Making of Birmingham*, p. 142: "the new establishment, when completed, being known (from a wayside inn in the neighbourhood called 'the Soho') as the Soho Factory".

from London to Birmingham and fathered on the house in question. Of course there is always the possibility that the name was borrowed directly from London just as, for example, were Vauxhall in Birmingham, Piccadilly in Manchester and Charing Cross in Glasgow. It is a matter for remark that the older Birmingham inhabitants always referred to the place as "The Soho". Obviously we have here a problem that admits of, and deserves, further investigation.

The memorandum from which we have quoted shows that Boulton made considerable improvements in the buildings, and in fact the Manufactory that we see represented in so many views of the period was an entirely new building of some architectural pretensions with a bold elevation on three floors planned on a generous scale. One of the earliest and best of these views of the Manufactory is that shown in our frontispiece; its date is not known but internal evidence shows it to be *c.* 1780. The entrance with a clock tower was in the middle and on either side were wings forming a large open yard at the back on lower ground. The upper rooms in the wings were devoted to dwellings for the workmen and their families—there are curtains in the windows, it will be noticed. It is not surprising to learn therefore that much thought and time were given to the scheme as is proved by another memorandum of Boulton:

> All the time M.B. spent in considering exp^{tg} & contriving ye Necessarys for ye s^{d} Mill & Business wch is there car^{d} on wch took him nearly one Year & a half before ye partnership commenced.

We know little of Boulton's partner, John Fothergill, beyond what is disclosed in the memorandum quoted. It is of course obvious that Boulton, being primarily a manufacturer, needed someone who was in the merchanting business with a connection in foreign markets. Fothergill was just such a person and it happened that he was at liberty, but he was not a man of any substance and he seems to have been unenterprising and rather a money-grubber. The date on which the partnership commenced was quarter day June 24th, 1762. It appears that each partner put £5000 into the business; for his share Fothergill borrowed the money. Boulton's clerks at Snow Hill, Zacchaeus Walker[1] and others, were occupied for eight days in taking stock there and preparing an inventory wherewith to start the partnership. Shortly after Midsummer the removal took place, but Snow Hill was retained as a warehouse and a dwelling. Boulton attended to the organization and production, while Fothergill devoted his attention mainly to establishing home and foreign agencies and for that purpose travelled extensively, visiting the principal trading centres in France and Germany.

Boulton did not at once go to live at Soho; it appears from a memorandum left by him that he allowed his mother to live there, but later, when Fothergill was in need of a house, Boulton offered it to, and it was accepted by, him. Much to the annoyance of Boulton, Fothergill neglected the place. His "love for mony was greater than for a garden"; the land had "not received ye assistance of one L^{d} of muck since F's residence". Boulton decided therefore to occupy the house himself, as he had originally intended. We assume that his mother returned to Snow Hill or rather Slaney Street when Fothergill went in, and stayed there till her death which occurred in 1785. Boulton was enabled to turn Fothergill out because the dwelling house, as will be seen from the memorandum first quoted, was not included in the partnership. There would be an additional reason for moving out to Soho, if this incident coincided, as it seems to have done, with Boulton's second marriage, for after Lichfield his wife might not care for the somewhat dingy condition into which Snow Hill, despite its name, was now falling.

At Soho, Boulton and Fothergill, while carrying on still the same kind of manufactures as the former had done at Snow Hill, were now able to enlarge greatly the scope of their products.

One of the industries the partners took up was the manufacture of steel jewellery, but how early we are unable to say. This jewellery is made of facetted steel, highly burnished; it was

[1] Son of Robert Walker (1709–1802), styled "Wonderful Walker" by the Cumbrian peasantry, curate and schoolmaster of Seathwaite, Cumberland, 1735, till his death.

PLATE II. WEDGWOOD CAMEOS IN BOULTON CUT STEEL FRAMES

Courtesy of the Victoria and Albert Museum

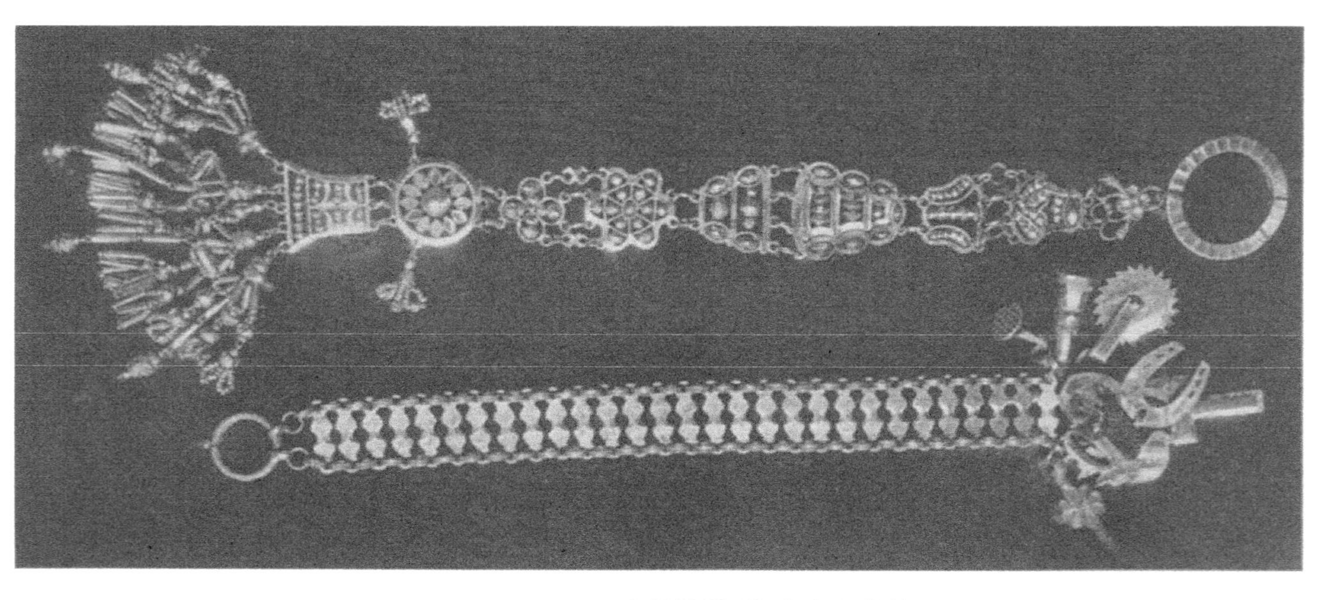

PLATE III. CUT STEEL FOB CHAINS

BOULTON AND WATT COLLECTION

Courtesy of J. D. Prior, Esq.

therefore resistant to rusting and reflected light brilliantly. This jewellery appears to have had its origin in England; a writer in 1690 encountering at Milan sword hilts, heads of canes and snuff boxes, decorated in this way, observes that "they can be had better and cheaper at Birmingham". It was there and at Wolverhampton that the industry was seated; watch-chains, chatelaines, buttons, buckles, clasps, purse-mounts and seals fashioned in this style were produced. Later, enamels from Battersea and Bilston, or cameos from Etruria were set in steel frames of this facetted steel. Our illustration (Pl. II) shows a setting of a Wedgwood cameo of the goddess Flora and another of the zodiac sign Libra, made by Boulton and Fothergill, now in the Victoria and Albert Museum. This steel jewellery remained in high favour till the last quarter of the eighteenth century. In the first quarter of the nineteenth century taste and quality deteriorated greatly. The illustration (Pl. III)[1] of two watch-chains or fobs with their multiplicity of nick-nacks, is a proof of this; these chains are said to be Soho productions but we would like to disbelieve it, or at any rate to date them after Boulton's death. These watch-chains are accompanied by a pattern book of a very varied array of articles made in the same way. The production must have been enormous and it is surprising that not more have been preserved, e.g. in our Museums. Even Birmingham, the home of the industry, has very few to show. The demand for the jewellery declined and the last expiring phase of it was in the steel beads with which for example the slip-ring purses possessed by our great-grandmothers were decorated. The second quarter of the nineteenth century saw the final disappearance of this jewellery.

The steel from which such articles were fashioned was produced by the cementation process, already described, from wrought iron bars preferably of Swedish provenance. It was usual for a manufacturer to have a furnace of this kind on his premises and Soho was no exception. In the case of manufacturers of cutlery and edge tools the practice was invariable. How important a part was played by the furnace in the hardware

[1] B. and W. Coll. Courtesy of the lender, Mr J. D. Prior.

trade is evidenced by the fact that a street in Birmingham—Steelhouse Lane—is named after a furnace that existed there in former times.

Another manufacture that Boulton and Fothergill took up—it may even have been carried on at Snow Hill—was that of buttons. At all times an important industry in Birmingham, it was then particularly so as the button was an essential article of dress—almost as much so as the pearlie to the costermonger to-day—and was made in almost inconceivable variety. We can best describe this in the words of the *Directory*[1] already quoted:

Button Makers.

This Branch is very Extensive, and is distinguished under the following Heads viz. Gilt, Plated, Silvered, Lacquered and Pinchback, the beautiful New Manufacture Platina, Inlaid, Glass, Horn, Ivory, and Pearl: Metal Buttons, such as Bath Hard and Soft White, &c. there is likewise made, Link Buttons in most of the above Metals, as well as of Paste, Stones, &c., in short the vast Variety of sorts in both Branches is really amazing....

From the foregoing we should infer, and rightly so, that the appearance and the material, not the utility of the button, were the desiderata. We are not surprised to learn in consequence that the price varied enormously. In 1781 buttons were to be had at any price between 3*d*. and 140 guineas a gross. From this and the description it follows that the range of materials, metals and alloys drawn upon to make buttons and the consequent variety of technique required in their manipulation was extensive and continued to expand, not only to meet the demands of fashion, but also owing to the efforts of the manufacturers themselves to stimulate the craze for novelty.

It is not surprising that with this expansion of activities—the partners were now employing about 600 work-people—extensions to the Manufactory were necessary. By 1765 they had already expended £4000 on buildings and now they had to spend half as much again. What the position was, how their business appeared to an outsider, and how Boulton received an offer to go abroad, is reflected in correspondence that took place

[1] Sketchley, *loc. cit.* 1767.

in this year between the second Earl of Halifax, Secretary of State, and Dr Roebuck. This correspondence is preserved in the Record Office.[1]

About a year ago you applied to the Earl of Halifax to stop some artificers who were going to Sweden to establish a manufactory of Iron and steel in that country, which his Lordship effectually did for that time but his Lordship has lately received advices that such a project is now on foot again and that one Dr Solander[2] who is thoroughly acquainted with the affair is employed to negociate the business and there are reasons to believe he has been at Birmingham with proposals to one Boulton who the Sweeds are very desirous to engage to leave this country and settle amongst them.

Mr Stanhope then asks Roebuck to act practically as a spy:

It would be particularly usefull to his Lordship to know if Dr Solander has been at Birmingham, or Boulton in London, or absent for any time within a few months past.

Mr Stanhope, who was private secretary to Lord Halifax, reminds Dr Roebuck of the Act 5 Geo. I, cap. 27; this, enacted to stop the practice of enticing artisans to go abroad, imposed a fine not exceeding £100 and 3 months' imprisonment for the offence.

James Farquaharson, clerk to Roebuck and Garbett, replied[3] that he had authority to open letters to Dr Roebuck who is "in Scotland & will be for some months to come" and said:

I have made all the secret inquiry I can to learn if Dr Solander hath been in Birmingham this year but I dont find that he has been...but I find about 2 years agoe he recommended 2 or 3 Swedish Gent[n] from London to come to see the Manufactures of this place.

Mr Boulton is a man of very considerable property in & about Birmgham & I think very unlikely to attempt going to Sweden. I am very well acquainted with him & within this 4 years he has laid out above £4000 in building Shops Mils & Utensils for manufac-

[1] Record Office: S.P. Dom. Entry Book 141, p. 18. Copy of letter from Mr L. Stanhope to Mr John Roebuck, 1765, 3 July.

[2] S.P. 37, 4, No. 33, 1765, July 6.

[3] Dr D. C. Solander (1736–82) was a Swedish botanist who accompanied Sir Joseph Banks on his expedition to the Pacific, 1768–71. He was known personally to Boulton and therefore a likely intermediary.

turing of buttons & Steel Toys & this present year he is building an addition to his work shops that will at least cost £2000 more. About 2 years since he took in a partner & their firm now is Boulton & Fothergill. If I can hear of any other project of this nature I shall certainly inform you & if I can be of any farther service you may depend I shall do it with the outmost secresy.

Mr Boulton hath not been from home much this year he was in the Whitsun week seeing the Duke of Bridgewater's works near Manchester.

Farquaharson must have interviewed Boulton on the matter to judge from a supplementary letter[1] which he addressed to Mr William Burke, as follows:

I can now inform you for certain that Mr Boulton has no intention of going to Sweden, tho' he has very advantageous offers made him. On the contrary he would spare no expense to detect any scheme of that nature which he considers so prejudicial to his country. Mr Boulton has lately rec[d] a letter from Doctor Solander wherein he says that he had wrote to him 4 times & that he imagines they have miscarried for want of a proper direction. The Doctor presses Mr Boulton for an interview, but he declines either writing to him or to meet him but he intends to send you copy or the original letter from Dr Solander....

If Mr Boulton hath not wrote you I can only impute it to his attention to his favorite scheme of increasing his Manufacture & overlooking his New Buildings; the latter (as I mention'd before) will cost him above £4000. Dr Solander hath not been in Birmgm that I can hear of.

To anyone who knew Boulton, such an offer, especially at this stage of his career, must have appeared ridiculous even had he no patriotic motives in refusing. Probably it appeared so to him for there is no evidence of his having written the intended letter.

The Act we have mentioned for preventing the enticement of artisans abroad was supplemented by another (23 Geo. II, cap. 13) which increased the penalties and, further, prohibited export of tools and industrial appliances. Needless to say, both

[1] *Loc. cit.* No. 52, 1765, Sept. 19.

these Acts were systematically evaded both by masters and by men.[1]

The reference to Boulton's visit to Worsley near Manchester to see the Duke of Bridgewater's Canal is significant because it shows that he was alive to the importance of canals; indeed he was a prime mover in their introduction only two years later into the Black Country.

Boulton, proud of his great Manufactory, was determined that not only should his products be of sound workmanship but also that they should be in the best possible taste. He wrote to Fothergill thus:

> The prejudice that Birmingham hath so justly established against itself, makes every fault conspicuous in all articles that have the least pretensions to taste. How can I expect the public to countenance rubbish from Soho while they can procure sound and perfect work from any other quarter?

In the pursuit of his ideal and with the very practical aim of finding fresh industries, cognate as much as possible to his own, that he could establish, he turned his attention to the manufacture of fusion or Sheffield plate, then springing up in that town. It is a testimony to his wideawakeness that he recognized so quickly the existence of an opening.

A word or two must first be said about the process and its origin. Thomas Boulsover (1740–88), a Sheffield cutler, in following his occupation of scale[2] maker, found that when a thin sheet of silver is fused or soldered to a thick one of copper, and the compound ingot or billet subjected to rolling or flattening, the two metals, and not the softer—silver—alone as might have been expected, extend equally; thus a material consisting of a stout sheet of copper with a very thin layer of silver is obtained. Boulsover appears to have been the first to discover, about 1742 or 1743, that the metals would behave in the manner indicated above. He was also the first to see somewhat dimly the com-

[1] There is much information on this subject in the Boulton Papers, cf. *Birmingham Jewellers' and Silversmiths' Association Monthly Bulletin*, Nov. 1932, p. 204.

[2] Scales are the pieces of horn, etc. on the sides of the handle of a knife.

mercial possibilities of the process. He applied the process to buttons which to all appearance were silver, wore like silver and he was able to sell them much under the price of silver. He kept the process secret for many years; it was Joseph Hancock who realized the wider possibilities of this plating process and about 1758 started making saucepans, coffee-pots, hot-water jugs, and candlesticks of the new material. He so far eclipsed Boulsover as to be considered in many quarters the originator of Sheffield plate. After the date last named, many other manufacturers took up the process.

It is easy to imagine that the enterprising Boulton would be one of the first to hear of the new process and it would at once occur to him that it was just the thing for his establishment; hence he determined to start the manufacture at Soho. There is a tradition that, sent by his father, young Boulton went to Sheffield, to acquaint himself with the technique of the process, to the workshops of Richard Morton, one of the ablest of the early makers. It is said further that there was some talk of a partnership between them. Taking into account the existing situation, it is very unlikely that there can be any truth in the story; what we can believe and what probably gave rise to the tradition is that Boulton went to Sheffield and perhaps stayed there some short time to master the process. If such were the case he did so to good effect, as is evidenced by the fact that he was successful in introducing the manufacture at Soho. The earliest mention of work done by him is in 1762, and for a quarter of a century after the initial discovery no factory for the production of Sheffield plate except that of Boulton existed outside Sheffield, indeed there was no other maker in Birmingham till 1770, when quite a number of manufacturers entered the field. Boulton's productions can be said to be fully equal to those of his Sheffield rivals, as an inspection of the Ellis Greenberg Collection, 1931, of Boulton plate in the Birmingham Art Gallery will establish.

Boulton did not introduce any technical improvement in Sheffield plate but he was one of the first to adopt the sterling "silver thread" edge of Samuel Roberts and George Cadman

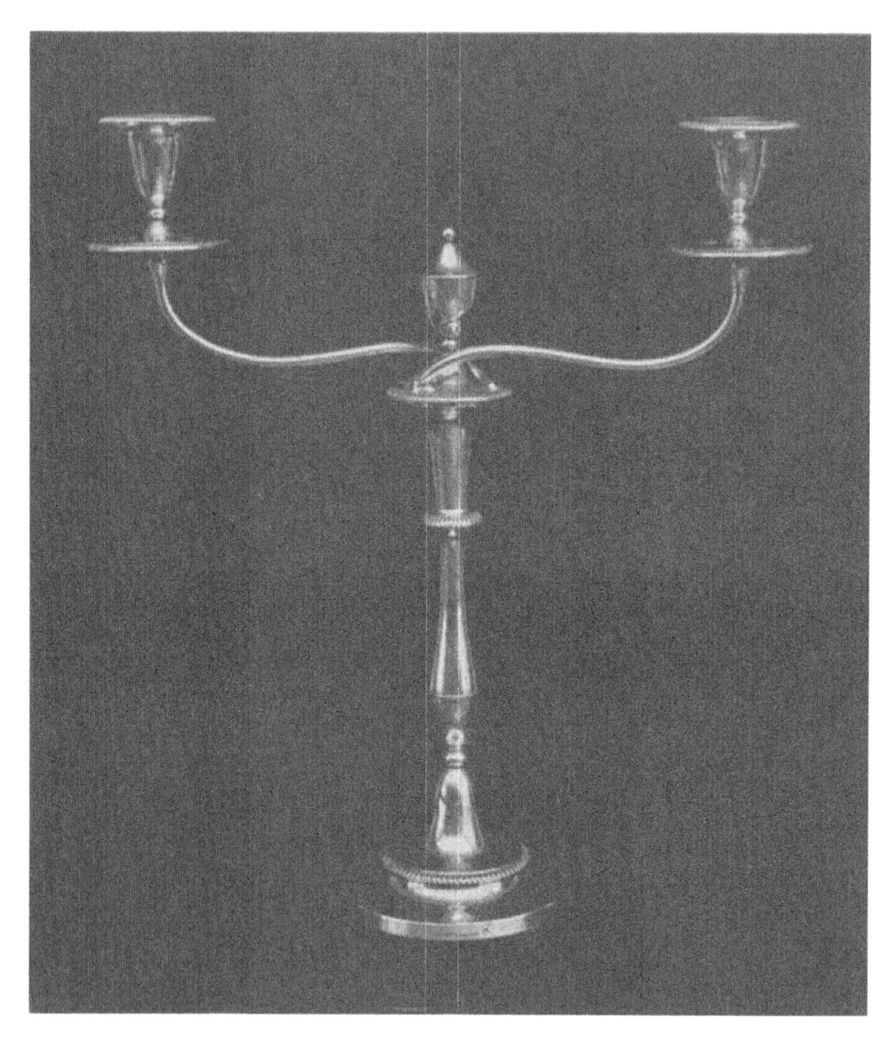

PLATE IV. SHEFFIELD PLATE CANDELABRUM, *c.* 1800

ELLIS GREENBERG COLLECTION

Published by permission of the Museum and Art Gallery
Committee of the Corporation of Birmingham

instead of the plated wire thread edge. From what has been said already, it will be realized that with the raw exposed edge of the plate thus covered with silver, where the greatest amount of abrasion takes place, no amount of cleaning or wear will disclose the copper. The Soho workmen brought the art of silver threading the edges to great perfection. Such ware Boulton stamped "silver borders" along with his mark of a sun registered in 1784. The sun is generally found struck twice, thus ☼ ☼, or else a single sun, thus ☼, with the word "BOULTON", is struck.

Every collection of Sheffield plate will be found to comprise some of Boulton's pieces not only because of their beauty of design but also because he was, we believe, the largest single manufacturer in the country. He excelled particularly in pieces with gadroon and shell borders, such as snuffer-trays, salvers, coasters (i.e. decanter-stands), tea-trays, coffee-biggins, candelabra and table-ware generally. Most of his marked articles are of late date, say from 1805 onwards. As one would expect from Boulton, they are of exquisite workmanship and of such good quality that they have suffered but little from the wear of a century and more.

Our illustration (Pl. IV) is taken from a candelabrum, one of a pair in the Ellis Greenberg Collection mentioned above; this tasteful object probably dates from 1800. If carefully inspected, it is possible to see on the plinth the words "silver borders" and a single sun, as already described.

Shortly after the Sheffield plate manufacture was established at Soho it occurred to Boulton to take up the making of silver plate, then greatly in vogue owing to the increase in wealthy families deriving their incomes from the new industry. It was only a short step from Sheffield to silver plate—the same designs could be used, the same workmen could be employed and no new selling agency was required. We shall learn subsequently that this manufacture was begun as early as 1765, but there were special difficulties in the way so that it was several years before anything serious was accomplished, and we therefore defer its consideration to a later page.

Another branch of artistic ware of which Boulton took up the manufacture was ormolu. In France *or moulu* was the name given to gold which had been ground up finely with mercury to form an amalgam used in gilding ornamental brass and other objects. In England ormolu is the name given to brass of high purity with a high content of zinc cast in ornamental forms and gilded, for the enrichment of other objects such as furniture. It was brought out first in France and had great vogue in the Louis XIV period. Fothergill if not Boulton would have seen examples in Paris during their visits there, and we know definitely that he was making it in 1767; probably he was doing so several years earlier if we are to trust a statement in the *English Gazetteer* of 1762 that speaks of Birmingham ormolu as being highly esteemed in Europe. This must refer to Boulton because at the time he took it up it was not being made elsewhere in England.

For the purpose of obtaining designs for his products Boulton sought the aid of his friends in order through them to borrow works of art to act as models. It is more than probable that these introductions were effected in the first instance through the Hon. Elizabeth Montagu (1720–1800) the leader of a salon in London, an authoress, and the first "blue-stocking"; she was by birth a Robinson and Boulton could claim kinship with her through his wife. Her assemblies were a focus of intellect and fashion for the metropolis. It is known that she admired Boulton's efforts to raise the standard of taste and to introduce art industries new to the country. In a letter to him she remarks:

> I take greater pleasure in our victories over the French in our contention of arts than of arms. The achievement of Soho instead of making widows and orphans make marriages and christenings....Go on then, sir, to triumph over the French in taste and to embellish your country with useful inventions and elegant productions.

It was through Mrs Montagu we assume that Boulton obtained introductions, as he did, to art patrons and connoisseurs who were of the greatest help to him. Such were the third Duke of Richmond and Lennox, the first Duke of Northumberland, the second Earl of Shelburne (afterwards the first Marquis of Lansdowne) and Horace Walpole. Through these he was able to borrow or make drawings of vases, candelabra, statuary and other works in metal. Boulton is said to have frequented the British Museum with the latter aim in view, but this is highly improbable, unless he was granted special privileges, as at that date the public were only admitted to the Museum on written application; not more than five persons were allowed in at a time and then only for about an hour and it is on record that the waiting list was quite a long one. Boulton frequented sale rooms for the purpose of acquiring art objects. Furthermore he employed artists to make designs for him; he did not confine himself to this country for on the Continent besides Fothergill, who was on the look-out for objects while on his business journeys, Boulton employed a man named Wendler on a mission to Italy to purchase art objects for him. What the nature of the commission entrusted to Wendler was, and what were the new manufactures that he had taken up, may be gathered from the following letter to him (1767, July):

> If, in the course of your future travelling you can pick up for me any metallic ores or fossil substances, or any other curious natural productions, I should be much obliged to you, as I am fond of all those things that have a tendency to improve my knowledge in mechanical arts, in which my manufactory will every year become more and more general, and therefore wish to know the taste, the fashions, the toys, both useful and ornamental, the implements, vessels &c. that prevail in different parts of Europe, as I should be glad to work for all Europe in all things that they may have occasion for—gold, silver, copper, plated, gilt, pinchbeck, steel, platina, tortoise shell or anything else that may become an article of general demand. I have lately begun to make snuff boxes, instrument cases, tooth picks &c. in metal gilt, and in tortoiseshell inlaid, likewise gilt and pinchbeck watch chains. We are now being completely fixed at Soho, and when Mr. Fothergill returns (which will not be for six months) I shall then have more time to attend to improvements than I have at present.

Such were some of the directions in which Boulton was continually feeling his way, and in many of them he did expand. At one time he even thought of engaging in the fictile industry.

Writing to his friend Wedgwood, at the time the latter was starting his new works at Etruria, Boulton said he "almost wished to be a potter". However, it did not get as far as that; he was satisfied with mounting in metal some of Wedgwood's productions. Boulton wrote (1769, Jan.): "The mounting of vases is a large field for fancy in which I shall indulge as I perceive it possible to convert even an ugly vessel into a beautiful vase". How successful Boulton was may be judged by his mounting of the cameos (see Pl. II) already mentioned.

Through his aristocratic friends Boulton obtained audiences of the King, George III and of his Queen Sophia. Boulton describes his visit to the Palace in a letter to his wife written, to judge from internal evidence, in 1770. The letter runs:

> The King hath bought a pair of cassolets, a Titus, a Venus clock and some other things and enquired this morning how yesterday's sale went. I shall see him again I believe. I was with them, the queen and all the children, between two and three hours....Never was man so complimented as I have been; but I find that compliments don't make fat nor fill the pocket. The queen showed me her last child which is a beauty but none of 'em are equal to the General of Soho or the Fair Maid of the Mill. God bless them both and kiss them for me.

These were his pet names for his two children Matthew Robinson, "Matt", and Anne, "Nancy".

The enquiry of the King as to "how yesterday's sale went" alludes to the fact that Boulton had adopted the practice of bringing his art objects to London and selling them at Christie and Ansell's. A printed invitation dated May 18, 1778, to attend an exhibition and sale of "Or Molu Ornaments from the works at Soho" in "Messrs Christie and Ansell's Rooms in Pall Mall" on May 20th, has been preserved.

In a subsequent gossiping letter he describes to his wife another audience of the King and Queen in these words:

> I am to wait upon their majesties again so soon as our Tripod Tea Kitchen (tea service, in plate) arrives and again upon some other business. The queen I think is much improved in her person, and

PLATE V. ORMOLU AND BLUE JOHN CANDELABRUM AT WINDSOR CASTLE, *c.* 1770

Reproduced by gracious permission of His Majesty the King

she now speaks English like an English lady. She draws very finely, is a great musician, and works with her needle better than Mrs. Betty. However, without joke, she is extremely sensible, very affable, and a great patroness of English manufactures. Of this she gave me a particular instance; for after the King and she had talked to me for nearly three hours, they withdrew, and then the queen sent for me into her boudoir showed me her chimney piece and asked me how many vases it would take to furnish it. "For" said she "all that china shall be taken away." She also desired that I would fetch her the two finest steel chains I could make....

Seeing that the Queen had come to this country from Mecklenburg in 1761 there was every reason why she should have been able to speak English by this time. The King and Queen were engaged in refurnishing some of their apartments at Windsor Castle. Much of Boulton's ormolu work found its way thither and to Buckingham Palace and many objects are still preserved there. We are fortunate in being able to reproduce an illustration of an important piece (see Plate V) from Windsor Castle and it may be taken as typical of Boulton's best work. The object is a candelabrum, one of a pair of elegant design showing Adam influence. A rich effect is obtained by making the centre vase of polished "blue john", a kind of fluor spar found in the Peak district of Derbyshire; the mine known as the Blue John mine, whence the spar was extracted, was discovered in 1743.

A pair of candelabra, closely similar and quite as elegant, is in the collections of the Victoria and Albert Museum. In this case there are three double-branched candlesticks rising from a vase of blue john, supported by three male figures standing on a triangular plinth which is mounted with richly chased trophies of arms; this in turn rests on a plinth of white marble. These candelabra were purchased at the Zetland sale in 1934. The Empress Catherine of Russia purchased similar objects for the Imperial Palaces and thought them superior in every way to French productions of the same kind.

The mention of a Venus clock in the letter to his wife reminds us that Boulton and Fothergill took up the manufacture of clocks—a business that one would have thought eminently within the capacity of Soho. At first Boulton's idea was to rival the ormolu productions of the French makers. To assist him in the clockwork mechanism he obtained the advice of John Whitehurst (1713–88) of Derby. The correspondence with the latter on this subject dates between 1769 and 1772. Boulton did make two highly finished clocks which he took up to Christie's salerooms in London but they did not find purchasers. In disgust he wrote to his wife as follows:

I find philosophy at a very low ebb in London and I have therefore brought back my two fine clocks which I will send to a market where common sense is not out of fashion. If I had made the clocks play jigs upon bells and a dancing bear keeping time, or if I had made a horse race upon their faces I believe they would have had better bidders. I shall therefore bring them back to Soho and some time this Summer will send them to the Empress of Russia who, I believe, would be glad of them.

For about a twelvemonth the clocks went begging and eventually, as there seemed to be no hope of finding a customer for them, they were sent out to Russia to the Empress through the intermediary of Earl Cathcart as a present and accepted by her.

Boulton now seems to have envisaged mass production of cheap clocks which one would have thought even at that time to have offered a lucrative field and one that he was well equipped to enter. He was not the first to have such a conception, for John Gimblett[1] already had about 1765 established a watch-making factory in Snow Hill, the first place in Birmingham or indeed in England where watches were made under factory conditions; this business however failed. However that may be, the clock-making business was suspended about 1772 and never resumed. James Keir, about whom we shall have much to say presently, a great friend of Boulton, wrote a memoir of Boulton in 1809 and from it we extract the following account of the horological business:

Among the various branches undertaken in this temple of useful

[1] Gimblett made also inlaid tortoiseshell wares (piqué), a business purchased by Boulton about 1780.

and elegant Vulcanian arts, none shewed more the extent & versatility of the mechanical & scientific talents employed at Soho, than the manufacture of *clocks*. It was always in Mr B's mind to convert such trades as were usually carried on by individuals into Great Manufactures by the help of machinery, which might enable the articles to be made with greater precision & cheaper than those commonly sold. No article was better suited to the display of invention, and accordingly clocks were made of various constructions; of these, one contained but a single wheel, and a timepiece of this kind was made which proved the truth of the principles of the invention. This one-wheeled clock was invented by Mr B's intimate Friend Dr Small who added to his many excellent endowments inventive genius united with the more rare advantage of calculating the precise results of mechanical problems by his superior mathematical skill....

Mr B's occupations prevented the manufacture of Clocks from being put to a fair trial of sale. It was suspended to a more favorable opportunity which has not hitherto occurred.

What Dr Small's one-wheeled clock was like we should much like to know; it must have been a mechanical curiosity. He took out a patent in 1773, July 22 [No. 1048], for improvements in clocks but there is nothing in the specification about a one-wheeled clock, although it contained some ideas quite as startling such as dispensing with the hands of the clock and hydraulic governing instead of using an escapement. He coupled these ideas with the statement that his clocks were "of much more simplicity than have hitherto been in use". If Dr Small had anything to do with the project of making cheap clocks at Soho, his death in 1775 and the fact that Boulton's hands were more than full with the steam engine, as we shall soon see, put an end to the project.

We have mentioned above that Boulton's ormolu pieces exhibit Adam influence. How this came about is easy to explain. The brothers Robert and James Adam were at this time at the height of their fame as architects, furnishers and decorators in the style based on classical models but peculiarly their own. In carrying out their commissions—and their practice covered the finest mansions then being built in the kingdom—they were in constant need of artists to carry out their designs or work out details. In Boulton they found evidently a kindred spirit ready to work in harness. He must have carried out much work for them in ornamental hardware and ormolu the identity of which is now lost.

In 1770 (Aug. 14) we find James Adam (1728–94) suggesting to Boulton the manufacture of plate and ormolu in an elegant and superior style. Boulton in a considered reply (1770, Oct. 1, Letter Book, 1768–73) goes into detail as to what are his resources, his plant and his methods; he gives also his ideas as to the establishment of a sale-room in or close to the Adams's premises in Durham Yard, St James's. The letter is so informative as to what was going on at Soho at this date that we venture to give it *in extenso.*

Your favr. of the 14th Augt. I received and ought to have given you an answer before now but having very little Leisure and there being something in your Lr. that struck me as a matter of Consequence, I chose to turn it over a little in my mind before I made you a reply. I think with you and your Brother that an emense Manufacture might be established by improving the designs and taste of plate and other ornamental furniture in or Moulu and am convinced that no men have greater power than your selves. I like wise know from experience that such a manufacture may be conducted upon much more advantageous terms in a situation like mine than in London, I having seven or eight hundred persons employed in almost all those Arts that are aplicable to the manufacturing of all the metals, the simi metals and various combinations of them, also Tortois shell, Stones, glass, Enamels, &c., &c.

I have almost every machine that is aplicable to those Arts I have two Water mills employd in rolling, pollishing, grinding & turning various sorts of Laths. I have traind up many and am training up more young plain Country Lads, all of which that betray any genius are taught to draw, from whom I derive many advantages that are not to be found in any manufacture that is or can be establishd in a great & Debauchd Capital.

I have likewise establish'd a Correspondence in almost every mercantile Town in Europe Which regurlary supplies me with orders for the grosser & Current Articles which enables me constantly to employ such a number of hands as yields a Choice of Artists for the finer Branches and am thus enabled to errect and employ a more extensive & more convenient aperatus than would be prudent to errect for the finer Articles only—I have Long been convinced that the Shopkeepers in town are the Bane of all improvments as it becomes an imprudent thing for a man to make improvments which are attended with expence when he cannot reap the fruits of his own Labor, and that he cannot do when it passes through the hands of such a race of disingenuous persons as most of 'em are—I must therefore say that a proper connection in London and a commodious situation for Sale wou'd be a very agreeable thing to me: I think no situation superior to the neighbourhood of Durham Yard but my Ideas of a Shop or sale room are very different from yours for I wou'd rather choose a large elegant room up Stairs without any other window than a sky light; by this sort of concealment you excite curiosity, more, you preserve your improvements from Street walking pirates: The Nobility wou'd like that less publick repository. The Novelty wou'd please more and last longer than the present Mode of Exhibition—The great customers for plate are such as are not to be caught by Shew as they walk along the Street and you know that unprivate Shops are only Customary in London, for at Paris all their finest Shops are upstairs. If a large Room upon this plan cou'd be had with proper apendages in the neighbourhood of Durham yard, I shou'd be glad to become Tenant of it (whether we can fix upon any joint plan or not). The lower parts might be aproprieted to the sale of the lesser Articles of our Manufacture and for the reception of Gent[n] Serv[ts], the upper handsom room for plate d'or Moulu and such other fine toys as we make. You cannot but be Sensible that a vast variety of Articles may be comprehended under these heads. But when I consider that I have expended already more than Ten thousand pounds in the necessary buildings for my manufacture and that those Buildings are furnished with expensive Machines, tools & Materials and that a Large sum is necessary to be employ'd in the circulation of my ordinary Trade and that to execute any thing considerable in the way you mention, a very large Capital wou'd be required and besides this an able & Active partner to reside in London, I perceive Difficulties which can be obviated only by its being agreeable to any of your family or friends to add his money and time to what I cou'd employ on this occasion. I am obliged to you for the Sketches of Chains for Ld. Shellburne's Lamps which I shall avail myself of in a manner that I hope will be agreeable to you and his Lordship. I have already shewn them to him and shall proceed according to his directions which I had the honor of receiving a few days ago at Soho. I am glad that you have had our Engine in View in the construction of yours but will wa[i]ve entering upon that Subject now as I shall be much more able to do it when I have the pleasure of seeing you in town.

I am with the greatest regard Sr
Yr most Obedt hum st
M. BOULTON.

There are many points of interest in this letter. The new articles that were continually being called for demanded workmen with aptitude in accommodating themselves to the work. Evidently the old form of training by apprenticeship did not altogether meet Boulton's requirements, although it had done so in his father's day at Snow Hill. The result was that Boulton now favoured taking in "plain Country Lads" and giving them intensive training to find out and foster their special abilities. One thing he did set his face against was premium apprenticeship although he was frequently offered considerable sums to take such apprentices. No doubt these country lads were the "fatherless children, parish apprentices and hospital boys" of whom he speaks elsewhere and for whose accommodation he "built and furnished a house". The taking of such children into industry was greatly in favour at that period and led to gross abuses. We have been unable to trace any such to Boulton's door: what has been said suggests rather that they were well-treated. The policy of keeping the toy business at high pressure in order to provide a force of highly trained craftsmen who could be drawn upon for less regular work of the highest class, shows advanced ideas in organization.

James Adam in his reply to this letter (1770, Nov. 5) offered Boulton part of the brothers' premises in Durham Yard as a place "exactly fitted for such a purpose", i.e. a sale-room. Boulton toyed with the idea but eventually abandoned it; it would have entailed keeping a stock of plate, etc. in London that would have locked up more capital than he and his partner could spare from the business, and apparently the Adams brothers did not see their way to entertain Boulton's suggestion that they should supply some of the capital themselves. It is possible that Wedgwood, the close friend of Boulton, may have suggested the idea of a showroom, because he established one in London himself and his successors have maintained such premises up to the present day.

In the important letter quoted above it will be noticed that Boulton says that all his apprentices "that betray any genius are taught to draw"; this again shows how farseeing he was. Where a product is repeated for hundreds of years, design becomes traditional—the worker and artist are one—but when new industries such as toy-making spring up, and there is a feverish search for novelties in them and in older ones, designing advances to an important function of industry. We do not learn who were the instructors of the apprentices or whether there were any persons employed at Soho as designers only. For giving instruction in drawing Boulton may have been taking advantage of existing facilities. In their evidence in support of a petition to the House of Commons praying for relief from the necessity to take out licences for dealing in silver when only small quantities are used as in the Birmingham toy trade, John Taylor and Samuel Garbett in their evidence said[1]: "That there are Two or Three Drawing Schools established in *Birmingham*, for the Instruction of Youth in the Arts of Designing and Drawing, and 30 or 40 *Frenchmen* or *Germans* are constantly employed in Drawing and Designing". This must be a very early instance of the establishment of schools of industrial art, and it does credit to the town, for nowhere was instruction more necessary; we should not be surprised if we were to find that Boulton had a finger in the establishment of these schools.

We have mentioned above that Boulton's success with Sheffield plate had turned his thoughts very naturally, as early as 1765, to the cognate industry of making silver plate. As soon as he began making solid silver articles, however, he found himself faced by a serious obstacle, for such articles to be vendible must be hall-marked and the nearest Assay office was at Chester, 72 miles away, or failing that, either London, 112 miles or York, 125 miles away. The cost of sending articles to any of these places, the likelihood of damage to them, the risk

[1] *Journn. H. of Commons*, XXVIII (1759), p. 496.

of highway robbery, the delay involved, and the possibility of designs being copied, were at once manifest.

In May, 1766, the Earl of Shelburne, who by the way considered Boulton "the most enterprising man in differt ways in Birmingham", with his Countess made a four days' stay in Birmingham; they visited the principal manufactories in the town, including, of course, that of Boulton. On their return to Bowood Park, the Earl penned an account of what they had seen and he makes this interesting observation, no doubt based on something that had dropped from Boulton's lips:

> Another thing they are in a great way of is an assay master which is allowed at Chester and York; but it is very hard on a manufacturer to be obliged to send every piece of plate to Chester to be marked.... It would be of infinite public advantage if silver plate came to be manufactured here as watches lately are,[1] and that it should be taken out of the imposing monopoly of London.

The matter was ever present to Boulton's mind and he hunted out the information that the faculty of hall-marking had been conferred on the goldsmiths of London by the appointment of a resident Surveyor or Assay Master under the first Assay Act in 1300. Provincial goldsmiths were under the disability of sending their goods to London to be hall-marked. Subsequently the goldsmiths of certain provincial cities agitated for and were granted the same privilege as London by the appointment of Assay Masters in those cities: viz. York, Exeter, Bristol, Chester and Norwich, by Act of Parliament 12 and 13 William III, cap. 4, and Newcastle-on-Tyne by 1 Anne, cap. 9. Thus there was, Boulton found, every reason to believe that, where the need arose, Parliament would sanction the setting-up of Assay Offices in other places.

An incident occurred in 1771 in connection with an order from Lord Shelburne for two silver candlesticks that was almost the last straw. Boulton, writing to his Lordship on January 7th, apologizes for delay in executing the order by saying that the candlesticks had been away twelve days at Chester to be marked

[1] The reference to watches is to Gimblett's factory in Snow Hill, already referred to, p. 58.

and that they were so carelessly packed at Chester for their return journey that the "chasing was entirely destroyed", that "we have been obliged to substitute new parts" and that "it will take near a week's work to make the necessary repairs". He goes on pregnantly to say:

I am so exceedingly vex'd about the disappointment and loss which have attended the two pairs of candlesticks that altho' I am very desirous of becoming *a great Silversmith,* yet I am determined never to take up that branch in the Large Way I intended *unless powers can be obtained to have a Marking Hall at Birmingham.* This is not the first time by several that I have been served so. I had one parcel of Candlesticks quite broke by their careless packing.

Having determined to appeal to Parliament, Boulton prepared the ground with great care. In 1772 the proposed Assay Office was discussed publicly, as is shown by passages in a letter (1772, Dec. 4), from Boulton to the Duke of Richmond:

I am now manufacturing some plate with a degree of Elegance w'ch I flatter myself your Grace would approve, and if Parliament would but grant us at Birmingham the same indulgence as it hath granted to many towns in England...I think I should push that article into a cheaper and more elegant style than tis now in, and for that purpose I expect the Town of Birmingham will present a petition to Parliament this session praying for the establishment of an Office here, for the assaying and marking of wrought plate.

The matter got into the public press, as no doubt Boulton intended, and the Sheffield manufacturers heard about it. Naturally they wanted to have a similar privilege and on December 8th, 1772, Mr Gilbert Dixon, attorney, Clerk to the Cutler's Company, wrote to Boulton and Fothergill, whom he supposed to be "the most principall persons in this Business" and said:

The Establishment of such an Office in Sheffield has been much wish'd for by the Artificers in Silver here, who for some time past have had such a project in contemplation for the convenience of having their Goods assayed and marked without the trouble, expence and delay of sending them to the Goldsmiths Hall in London.

Sheffield, although farther, 159 miles, from London than Birmingham, was much nearer an Assay Office, viz. York, 45 miles, and Chester, 65 miles, but distance was one of the least of the disabilities already enumerated so that for all practical purposes Birmingham and Sheffield were in the same boat. Mr Dixon suggested that the Sheffield people should "go hand in hand with you to Parliament" and pointed out, with true legal mind, a pitfall in that neither of these towns was incorporated and therefore there was no one before whom the intended Assay Master could be sworn.

Boulton, replying in the third person on Christmas Eve 1772, welcomed the support of Sheffield in these generous words: "he hath no selfish contracted views but most sincerely wishes to see Sheffield as well as Birmingham upon a footing with London in that point & likewise assures them that to the utmost of his power he will assist & promote their endeavours".

As a matter of fact Boulton's idea was, as we should expect, statesmanlike in that he wanted the Government to promote a Bill that would give power to establish assay offices in *any* city or town where the manufacture of plate would justify it and on the other hand to close such of the existing offices as did not satisfy this criterion,[1] but he realized that this would raise a bigger storm of opposition than the more modest scheme would do; accordingly the latter alone was proceeded with.

Sheffield artificers presented their petition to the House on February 1st, 1773, and Birmingham on the following day. The fact that Sheffield's petition was in first has led some persons to believe that the agitation originated there, and not with Birmingham. Both petitions were referred to a Committee of the House to be examined. For the information of Members a "Memorial relative to Assaying and marking wrought plate at Birmingham" signed by Boulton was circulated by him. This is such a masterly and logical document that we would fain have quoted the whole of it had space permitted. We cannot, however, resist quoting the last paragraph as it is so thoroughly Boultonian:

[1] It is of interest to know that the Assay Offices at Bristol, Exeter, Newcastle-upon-Tyne, Norwich and York have been closed since Boulton's time.

Objections may possibly be made by the Corporations of Goldsmiths in *London* and in the other Marking Towns to such a grant, which, it may be said, may prove injurious to them, because it may enable others to work in Plate with as much Convenience as they now do. But as *Birmingham* is not near to any Market for Plate, it can deprive the other Towns of no Part of their Trade except by working better than they do and cheaper; and against losses of business by these means the proper Securities are not Privileges, but Excellence in Design and Workmanship and moderate Prices.

As Boulton expected, "objections" were not long in being heard, indeed they were already on the way. On February 17th, the Wardens and Assistants of the Company or Mystery of Goldsmiths of the City of London put in a petition, quite a masterpiece of obstruction, concluding: "THEREFORE it is humbly submitted that the Legislature will find it improper and dangerous to establish an Assay Office for assaying and marking of Gold and Silver Plate in the said Towns of Sheffield and Birmingham or each of them".

On the same day the "Goldsmiths Silversmiths and Plateworkers of the city of London" also sent in a petition against the Bill. The gist of this petition was that everything was for the best in the best of all places—London—and consequently that nothing should be done to disturb the hallowed procedure. The petition has about it a very familiar ring, in fact the arguments advanced were the ones that have done duty since the year one when change of any kind is proposed.

These petitions were referred to the same Committee that was dealing with the petitions for the Bill. This Committee sat on February 18th and subsequent dates. The only witness called from Birmingham was Mr Samuel Garbett. This was a stroke of generalship, for he could be considered a disinterested person. He disclosed some interesting facts, e.g. he said that he was a refiner of gold and silver, that Boulton and Fothergill began making silver plate about seven years prior to that date and made tureens, candlesticks, vases, coffee pots etc. and that they bought several thousand pounds worth of silver from him in a year; that there were upwards of forty licensed makers of plate in Birmingham. He said also that the greatest article of manufacture there was Buckles, employing more than 5000 work-people.

On February 25th a petition from Merchants and Manufacturers of Birmingham was presented alleging that they laboured under great disadvantages by not having an Assay Office in their town. The Goldsmiths, Silversmiths and Plate-workers of London countered it on the same day by a further petition, making a case out of "the deceits and frauds" with regard to plate practised by alleged bold bad men in Birmingham and Sheffield. The petitioners were of course intentionally clouding the issue because they were alluding to Sheffield plate. This brought countercharges from the other side, alleging that the London mark was put on silver below standard, i.e., what should have contained 11 oz. 2 dwt. silver in the lb., only contained 11 oz. This led to the appointment on February 26th, of a Special Committee to examine into these allegations. The sittings of the two committees then went on concurrently. On March 12th, leave was given to bring in the Bill; it was presented on March 26th and read for the first time.

It is impossible to go into all the details of further petitions, the evidence of witnesses, the debate on the Bill, the replies to petitions and the newspaper controversy[1] that went on. Suffice to say that the Assay Bill, with some amendments suggested by the report of the Committee which has examined into the alleged abuses, was ordered to be engrossed on May 18th; it was read a third time. On the 28th the Lords agreed to the Bill and the same day the Royal Assent was given.

Boulton had been indefatigable in lobbying and securing interest from all quarters. So successful was he that Dr Small on January 28th, 1773, ventured to pull his leg gently by writing: "I hope the King and royal family, the Nobility and the Ministry and your other friends are well". In another letter, 1773 (undated but docketed 11 May) to Fothergill, Boulton says:

[1] All this and much more is set down by Mr A. Westwood in *The Assay Office at Birmingham, Pt. I, Its Foundation*, 1936, to which we are indebted for most of the account here given.

I was in hopes that I should have been able by this post to have inform^{d} you of a final decision of ye House upon ye Assay Bull but as it did not b(r)eak up before eleven last night ye Members were so much fateagu^{d} that the House hath not sat to-day & our affair is now ajourn^{d} till Thursday so that you may expect to hear on Saturday what our fate will be. The Londoners have engaged two Councelors to plead their shabby cause at the Bar of the House & are makeing all possable Interest they can. However I do not dispair.

And as to the House of Lords I have twice the interest in that House than in ye Lower house. I shall not employ any Council but as I have taken great pains to make proper impressions upon ye Members I shall submitt ye case simply to their decision. Lord Denbeigh took me about with him yesterday in his Chariot to several ministerial Members pressing them to serve us. He says he has talk'd much to the King in my fav^{r}.

Lord Dartmouth had been of great support in the Lords and Sir George Savile in the Commons. Boulton had written in the letter already quoted May 11th, "I am damned sick of London but can't leave it at present". It was a tremendous relief to him when it was all over, and he could record in his Diary: "Set out from London 2 o'clock on Sunday ye 30 May & ariv'd at Soho ye Monday before 8 in ye morning,"—note how the speed of the coach had improved since 1747. He was welcomed home by the ringing of a peal on the bells of Handsworth Parish Church and he deserved it. "For it was a famous victory".

As was customary at the time, much of the interviewing and business of the Bills before Parliament was carried on in the inns and in the coffee houses that were such a feature of seventeenth and eighteenth century London. The "Crown and Anchor" in the Strand was easily first in this respect and was familiar to everyone who came up to town. It is but a conjecture, albeit a probable one, that the tavern sign emblems suggested themselves, perhaps to Boulton, as suitable assay marks. Whether this was their origin or not, and whether decided by toss of a coin or otherwise, the fact remains that in the Act of Parliament (13 Geo. III, cap. 52), setting up the two Assay Offices, the anchor mark was assigned to Birmingham and the crown to Sheffield.

The Assay Office in Birmingham was set up with James Jackson as the first Assay Master. He was sworn in by the Deputy Master of his Majesty's Mint and there was not the least difficulty about it. The first day the Office was opened, Boulton and Fothergill marked the occasion by sending in no less than 841 oz. to be assayed; one article alone weighed nearly 200 oz. and another over 100 oz. One of the pieces assayed in this year and the earliest to be preserved is now in the Assay Office, Birmingham; it is illustrated in Plate VI. The Assay Office mark is the anchor thus ⚓, preceded by the sovereign's mark, the lion passant if the article is of silver. The firm's mark was MB IF, but after Fothergill's death in 1782 Boulton used the mark MB alone.

It only remains to say that in 1784 the Act was amended to allow makers of plated ware to register their marks at the Sheffield Assay Office. Boulton registered the marks of two suns ✿ ✿ and BOULTON ✿, as has already been described on p. 53.

We have already shown that some of Boulton's ventures were not successes financially and this was particularly true of the more purely artistic ones of ormolu and, as we shall learn later, of pictures. At their best they must have been parasitic on the comparatively humdrum products of buckles and buttons. In the memoir by Keir from which we have quoted, this very prosaic reason for giving up making ormolu is dwelt upon. The passage in question is prolix but as it is so redolent of the approved style of composition of the eighteenth century we quote it at length:

Thus Mr B. gave free scope to his active genius in the introduction to Soho of various manufactures each of which would have been sufficient for the sole occupation of a mind of ordinary energy. But ingenuity is not sufficient to ensure profitable success in Trade. Many favorable circumstances must concur. A subordinate genius, an uniform perseverance, attention confined to one or few objects, a mind moved only by one passion, the love of gain, are qualities favorable to the acquisition of wealth. Genius may plant the Hesperian Tree, but patient dullness more frequently reaps the golden fruit.

PLATE VI. SILVER TUREEN, 1776–77

Courtesy of the Assay Office, Birmingham

Besides, local and many other circumstances must be favorable to every commercial undertaking without which no ingenuity & diligence can succeed. Many well contrived schemes fail, from some unforeseen and apparently minute obstacle. Accordingly, many of the above mentioned branches of manufacture on which so much ingenuity taste & capital were expended, did not make suitable returns of profit, but were rather rewarded with the fragrant odours of Praise & admiration, than with more solid advantage. It is even probable that they would have injured his capital too much, if their losses had not been abundantly compensated by the staple articles of less splendor in his manufactory....

Thus the or moulu vases & ornaments which gave models of good taste to other artists, were found to be too expensive for general demand, and therefore not a *proper* object of wholesale manufacture. Besides in all such articles of fancy every purchaser chuses to display his own taste (sometimes his want of taste) by chusing his designs & undervalues what every one may possess....

It was found that services of silver plate and other large silver articles could not be carried on with equal advantage anywhere as in the great market of London....

But if Mr B. did not receive from the or moulu & the other elegant branches of his manufacture the intended recompense of all commercial industry, it is certain that they greatly tended to his celebrity & admiration of his various talents, taste & enterprise. The desire of visiting Soho became a fashion among the higher & opulent ranks, foreigners of distinction & all who could gain access to it. He received his visitants with so much Courtesy & desire of pleasing (which were very distinguishing traits in his manner) that however agreeably their curiosity was gratified, they were still more pleased with the proprietor. This daily concourse of visitants was found to occupy too many of those hours which could not be spared from his own concerns for as *Knowledge* has been well said to be *Power* so *Time* in a man of business is *Wealth*. It was not without great difficulty & repeated instances of his friends that Mr B. could be persuaded to abridge or abandon this luxury of obliging. But however inconvenient it might be at the time to bestow so many hours in indulging the curiosity of others, he was not without a subsequent reward. He thus gained the acquaintance of most men distinguished for rank, influence & knowledge in the Kingdom, the good effects of which he felt in his several applications to Parliament, which were necessary to defend his property in the Steam Engine patent from invasion; and in his transactions of business with the ministers respecting coinage. Upon these occasions he met with the most ready & general support, accompanied with marks of esteem, goodwill & desire of making a return for the civilities he had shown at Soho.

The daily concourse of visitors to Soho referred to by Keir, was quite notable. In its early days the Manufactory was one of the show places of Birmingham and no distinguished visitor to the town failed to visit it. Our frontispiece with the travelling carriages of "the quality" in evidence emphasizes this. On Boulton's part it was sheer pride in the establishment but, as the memoir shows, it had valuable results in making him known to persons in high places, and in advertising the products of Soho. In showing the works to all the world and his wife, he was taking a modern view and one far removed from the practice of other manufacturers of his day who were secretive and uncommunicative to the last degree.

Not only did Boulton delight in showing people over the Manufactory, but he entertained them at Soho House, and so constant was the stream of visitors that it resembled an inn rather than a private house. To his friends it was "l'Hôtel de l'Amitié sur Handsworth Heath". In August 1767 he wrote to his London agent: "I had lords and ladies to wait on yesterday; I have French and Spaniards today; and tomorrow I shall have Germans, Russians and Norwegians". On another occasion he wrote:

> Last week we had Prince Poniatowski, nephew of the King of Poland, and the French, Danish, Sardinian and Dutch ambassadors; this week we have had Count Orloff and the five celebrated brothers who are such favourites with the Empress of Russia; and only yesterday I had the Viceroy of Ireland who dined with me. Scarcely a day passes without a visit from some distinguished personage.

He even had the Empress Catherine herself in 1776. With this succession of visitors—prince, diplomat, artist, author or merchant—one wonders how Boulton found time to do any work at all!

Among the authors we must mention the great Samuel Johnson. Very probably he knew Boulton through the Robinsons of Lichfield his native town. Johnson records his visit in his diary thus:

Tuesday Sept. 20th [1774]...we then went to Boulton's who with great civility, led us through his shops. I could not distinctly see his enginery. Twelve dozen of buttons for three shillings! Spoons struck at once!

Boswell records his visit two years later[1] and as it records the much-quoted dictum of Boulton about power—the commercial manufacture of the steam engine had just been started—we give it in full:

Mr Hector was so good as to accompany me to see the great works of Mr Bolton, at a place which he has called Soho, about two miles from Birmingham, which the very ingenious proprietor showed me himself to the best advantage. I wish that Johnson had been with us; for it was a scene which I should have been glad to contemplate by his light. The vastness and the contrivance of some of the machinery would have "matched his mighty mind". I shall never forget Mr Bolton's expression to me: "I sell here, Sir, what all the world desires to have,—Power". He had about seven hundred people at work. I contemplated him as an *iron chieftain*, and he seemed to be a father to his tribe. One of them came to him complaining grievously of his landlord for having distrained his goods. "Your landlord is in the right, Smith, (said Bolton). But I'll tell you what: find you a friend who will lay down one half of your rent, and I'll lay down the other half; and you shall have your goods again."

As time went on the practice of showing persons over the Manufactory was discontinued partly owing to the fact that the manufacturers agreed upon this step among themselves owing to alleged pirating of designs and information, and partly perhaps in Boulton's case because large quantities of the precious metals were being worked up on the premises; later on when the coinage for the Government was undertaken (see Chapter VII) this might be an added reason. At any rate at the close of the century, with the junior partners at the helm, no person on any consideration whatever was allowed over the works, as the following notice (B. & W. Coll.) witnesses:

[1] *Life of Dr Johnson*, 1791, ii, 32. The "Mr Hector" mentioned was Edmund Hector, an old schoolfellow of Johnson.

SOHO MANUFACTORY

THE PUBLIC are requested to observe that this MANUFACTORY cannot be SHEWN in consequence of any Application or Recommendation whatever. Motives, both of a public and private Nature have induced the Proprietors to adopt this Measure and they hope their Friends will spare them the painful Task of a Refusal.

SOHO, May 20 1802.

We have now far outrun the constable and must retrace our steps to tell the story of the introduction of Watt's steam engine, the most important service, in its ultimate effects on the world, performed by Boulton.

CHAPTER IV

BOULTON AND STEAM POWER

Extension of Manufactory—Shortage of water power—Boulton's steam engine—Watt's improvement of the steam engine—Roebuck's share—Watt visits Soho and meets Boulton—Watt's patent for the separate condenser—Partnership mooted—Roebuck bankrupt—Boulton acquires his share in the patent—Watt removes to Birmingham—Extension of patent—Partnership.

SOHO Manufactory had gone ahead not only because of the enlargement of the scope of the products but also because of the energy of the partners in securing orders from the Continent. It had gone ahead so fast that, as we have seen in the last chapter, extensions of the buildings were called for. This extension of premises brought with it the desirability of increasing the power as well. The supply of water from Hockley Brook, never too ample except in flood time, fell short seriously in summer. At first glance there seemed to be nothing available as an auxiliary power except the old horse mill, the very thing Boulton had gone out to Soho to avoid on account of its expense. But he was a man, as we have seen, who always kept himself well-informed as to what was going on in the larger world around him. Hence his mind turned to the expedient, already adopted by other persons, of installing a steam engine to lift the water from the tail race of the water wheel back to the pool or mill dam, in order that the water might be used over again. It is to be borne in mind that there were then no means of obtaining rotative motion by steam power.

There were two types of steam engine then available—the Savery engine like the pulsometer pump of the present day that raised water by direct steam pressure on the surface of the water, and the Newcomen or atmospheric engine which we shall describe shortly. It was the first named of these that Boulton considered would be suitable for his requirements. It appears that he had had a model of this engine made and had it sent to Benjamin Franklin whose opinion upon it Boulton begged for in the following letter (1766, Feb. 22):

My engagements since Christmas have not permitted me to make any further progress with my fire engine but as the thirsty season is approaching apace necessity will oblige me to set about it in good earnest. Query which of the steam valves do you like best? Is it better to introduce the jet of cold water at the bottom of the receiver or at the top?...My thoughts about the secondary or mechanical contrivance of the engine are too numerous to trouble you with in this letter and yet I have not been lucky enough to hit upon any that are objectionless. I therefore beg, if any thought occurs to your fertile genius which you think may be useful or preserve me from error in the execution of this engine you'll be so kind as to communicate it to me.

Franklin did not reply till Mar. 19 and then it was only to say that he had no suggestions to make beyond setting the grate of the boiler "in such a manner as to burn all your smoke". Easier said than done, for it is a problem that both engineers and laymen have struggled with ever since! Franklin concluded his letter by saying: "I sent the model last week, with your papers in it which I hope got safe to hand." Darwin apparently had heard about the model and wrote in his usual enthusiastic vein (1766, Mar. 11): "Your model of a steam engine, I am told, has gained so much approbation in London that I cannot but congratulate you on the mechanical fame you have acquired by it." The model was duly received at Soho and Boulton went on experimenting with it.

Now his friend Roebuck was about to start sinking coal pits at Bo'ness on the Firth of Forth and wished Boulton to join him as a partner to the extent of one tenth share. Boulton having his hands already full and it being a principle with him not to engage in any business which he could not oversee personally, very wisely, as it turned out afterwards, declined the offer. It was then that Roebuck told Boulton of a young man named James Watt, a protégé of Prof. Joseph Black, in business in Glasgow as a mathematical instrument maker, who had made an improvement upon the atmospheric steam engine, and had made a model of his invention. Naturally Boulton was greatly interested to hear of this novel engine and at once thought of it as a solution for his own difficulties; consequently he expressed a wish to see

Watt himself and hear about the invention as soon as an opportunity should present itself.

Watt's improvement consisted in effect in adding a separate

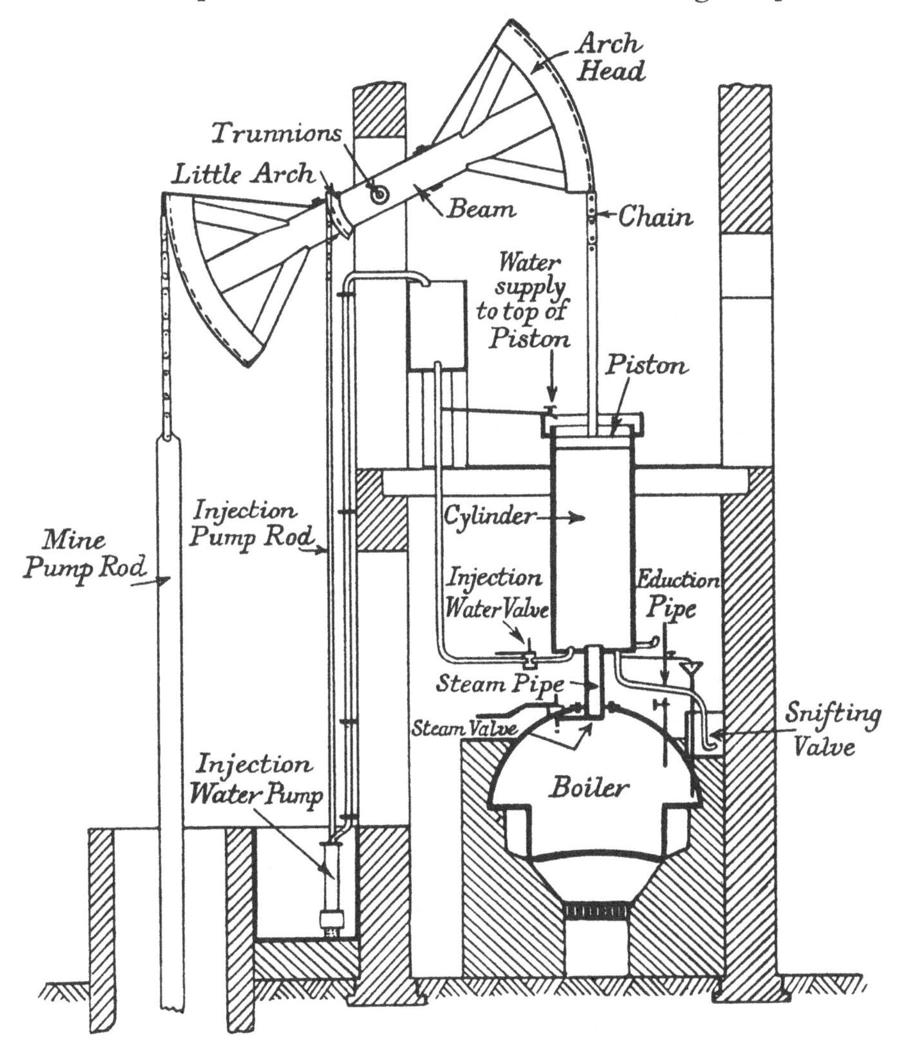

Fig. 2. Diagram of Newcomen's steam engine, 1712.
Courtesy of the Science Museum.

condenser to the atmospheric engine. This will be best understood by a reference to Fig. 2, which shows the engine as employed when Watt came on the scene. It will be observed that there was a piston working in an upright cylinder open at the top; the piston was attached by a chain to an equal-armed lever or beam—a giant pump handle in fact—to the other end of which were attached pump rods made heavier purposely than the piston to overbalance it and draw it to the top of its stroke, as shown in the diagram. If now steam—and it needed to be at no more than atmospheric pressure—was introduced into the cylinder under the piston, until all the air was blown out, and if then a jet of water was turned on so condensing the steam, a vacuum would be formed and the pressure of the atmosphere would force down the piston and consequently lift the pump rods with their load of water from the mine or other source.

It will be realized that the cylinder had to be heated from the temperature of the air to that of boiling water at every stroke and then cooled down again and this necessitated a heavy consumption of coal. Watt's great invention was to perform this condensation in a separate vessel, known as the condenser, and thus avoid the alternate heating and cooling of the cylinder. The effect of this was a saving of from two-thirds to three-quarters of the coal, as compared with the older engine doing the same work. It was not difficult for Watt to make a model, but to carry the idea into practice was quite a different matter for he had no experience in engine construction; thus the experimental work languished. It is hardly necessary to remark that a capital invention such as this is only brought to fruition by much thought, labour and expense.

There were other and more prosaic reasons for slowing down progress. Roebuck was sinking capital rapidly in his collieries and consequently had little money to spare for research: Watt had his living to earn, and to do so had set up as a surveyor and civil engineer. His services were so much in request that he had hardly any time for experiment. Roebuck might have suggested, but never did, maintaining Watt in order that the latter could give his whole time to the development of the engine.

The position was decidedly unsatisfactory, but a gleam of hope appeared. In connection with one of the undertakings for which Watt had made surveys—a canal between the Firth of

Forth and that of Clyde—he was called to London to give evidence for the Bill being promoted in Parliament for the scheme. For us this journey to London is of interest only from the fact that he travelled not by the Great North Road as he would normally have done but by way of the more westerly road through Stafford, etc. He branched off to Birmingham where he arrived on March 16, 1767, ostensibly to visit Samuel Garbett, but not finding him there, Watt went on via Oxford to London. On his return journey he travelled via Lichfield where he called on Dr Darwin to whom, under a pledge of secrecy, Watt confided his invention. On either the outward or return journey, he visited Soho, as he tells us himself (see Appendix I); Boulton was absent but Dr Small and Fothergill showed him the works.

During the following winter Watt did a considerable amount of experimental work on the engine and it was attended with such a measure of success that Roebuck was impressed so far that he desired to have a share in the invention. He agreed to take over the debt that Watt had incurred to Dr Black—amounting to about £1000—advance money for more experiments, and defray the cost of a patent—say £120—in return for a two-thirds share in the latter; Watt of course was to give his time and abilities. Agreement having been reached, Roebuck sent Watt to London to protect the invention by patent. Watt took the oath upon it on Aug. 9. On the 27th of the month he turned his face homewards, travelling via Lichfield and Stoke in order to be able to turn aside to Birmingham to visit Soho.

Now for the first time he met Boulton and at once the two men conceived a liking for one another. The older man, confident and business-like, saw in the younger one modesty and diffidence, it is true, but something more—ability of no mean order, only needing encouragement. The younger man saw in the older one a manufacturer who seemed to wield the wand of Prospero in the wonders he had achieved, as the previous short visit to the Manufactory had revealed. A longer time spent in the works with Boulton as cicerone more than confirmed the previous impression. The organization astonished Watt, the skill displayed and the beautiful work done fascinated the craftsman in him.

The stay at Soho House lengthened out to a fortnight and it was a time of unalloyed delight; the acquaintances too that he made were destined to become lifelong friends. The patent was a topic of conversation with Boulton and Small, the terms of the specification to be enrolled were discussed and a possible partnership was mooted. Watt could not commit himself of course till he had seen Dr Roebuck, but immediately after his arrival home, which was on October 11th, 1768, Watt called on Roebuck to tell him of Boulton's desire to engage in the scheme.

Now Boulton had already such a high reputation for business acumen that a favourable judgment on his part upon any scheme was an asset in its favour. So it proved with Roebuck for he at once urged Watt in a letter "much effectively to try the machine at large—a day, a moment ought not to be lost and you should not suffer your thoughts to be diverted by any other objects or even improvement of this, but only the speediest and most effectual manner of executing one of a proper size according to your present ideas". Splendid advice!—and salutary perhaps, for Watt could always see something further to try. The latter responded by sending during November drawings of an engine to be erected at Kinneil House, Roebuck's residence. Watt must have thought the advice would have been strengthened if accompanied by financial aid; as we shall see later, Roebuck gave him no such aid, beyond defraying the cost of the English patent.

This memorable patent for "a new method of lessening the Consumption of Steam and Fuel in Fire Engines" was sealed on January 9th, 1769 (No. 913). It covered England and Wales and Berwick-on-Tweed only—at that time separate patents for Scotland and Ireland were required. When drafting his specification, Watt had lengthy correspondence about it with Dr Small; he and Boulton advised Watt badly in this matter and as a result he committed two errors of judgment: firstly in omitting drawings, which he had actually prepared; and secondly in claiming a principle of action instead of the application of a principle. It was this second error that opened the door to attack later on when the validity of the patent was assailed in the law courts.

Roebuck was now more convinced than ever of the importance of the invention; as Watt said "the nearer it approaches to certainty he grows the more tenacious of it". Roebuck showed this by offering to Boulton no more than a beggarly licence to make the engine in the counties of Warwick, Stafford and Derby. Such a limited field had no attraction for Boulton—he wanted to "make for all the world". His letter (1769, Feb. 7) to Watt, declining this offer, is so remarkable for its grasp of the situation, its prescience and its enunciation of the principle of mass production, that we quote the whole of it:

> ...the plan (i.e. the Midland counties licence) proposed to me is so very different from that which I had conceived at the time I talked with you upon the subject that I cannot think it is a proper one for me to meddle with as I do not intend turning engineer. I was excited by two motives to offer you my assistance which were love of you and love of a money-getting ingenious project. I presumed that your engine would require money, very accurate workmanship and extensive correspondence to make it turn out to the best advantage and that the best means of keeping up the reputation and doing the invention justice would be to keep the executive part out of the hands of the multitude of empirical engineers who from ignorance, want of experience and want of necessary convenience, would affect the reputation of the invention. To remedy which and produce the most profit, my idea was to settle a manufactory near to my own by the side of our canal where I would erect all the conveniences necessary for the completion of the engines and from which manufactory we would serve all the world with engines of all sizes. By these means and your assistance we could engage and instruct some excellent workmen (with more excellent tools than would be worth any man's while to procure for one single engine) and could execute the invention 20 per cent cheaper than it would be otherwise executed and with as great a difference of accuracy as there is between the blacksmith and the mathematical instrument maker. It would not be worth my while to make for three counties only; but I find it well worth my while to make for all the world.
>
> What led me to drop the hint I did to you was the possessing an

idea that you wanted a midwife to ease you of your burthen and to introduce your brat into the world which I should not have thought of if I had known of your pre-engagement; but as I am determined never to embark on any trade that I have not the inspection of myself, and as my engagements here will not permit me to attend any business in Scotland and as the Doctor's engagements in Scotland will not I presume permit his attendance here, and as I am almost saturated with undertakings, I think I must conclude to.... No, you shall draw the conclusion; yet nevertheless let my conclusions be what they will, nothing will alter my inclinations for being concerned with you, or for rendering you all the service in my power and although there seem to be some obstruction to our partnership in the engine trade, yet I live in hope that you or I may hit upon some scheme or other that may associate us in this part of the world, which would render it still more agreable to me than it is, by the acquisition of such a neighbour.

What a statesmanlike letter! What bonhomie! What revelation of his own character! How patent the liking he had taken to Watt!

Although for the time being negotiations were thus brought to a full stop, the will on both sides to join forces still subsisted. To release Boulton from his difficulty with the water supply at Soho, Watt proposed that Boulton should "have liberty to erect one (i.e. engine) of any size for his own use" without payment. Watt supplied drawings for one early in 1770; an attempt was made to construct an engine but without success. And no wonder, for Watt himself, experimenting with the engine at Kinneil already mentioned, obtained only somewhat inconclusive results. Progress could have been vastly more rapid had Watt not been so engrossed with his civil engineering practice; he had really no time to devote to the engine although he kept corresponding with Small about it.

Meanwhile Boulton had to content himself ignominiously with the employment of horses to supplement his water power for turning the flatting mill, the grindstones and the laps for polishing steel ornaments.

The period 1772–3 is memorable as a time when severe trade depression followed by restriction of credit, in consequence of bad harvests and unwise speculation, spread over Great Britain. This is what Boulton enlarges upon in the following letter (1772, Nov. 10) to the Earl of Dartmouth then Colonial Secretary:

My Lord the trade of this country hath receiv'd a very severe shock since June last, even such a one as it will not recover [from] for many Years, particularly in Scotland; there is scarcely any person of considerable Trade in Great Britain but what hath felt the consequences of it in some respect or other either by the immediate effects of some of the great Numbers of Bankruptcies or from the general distrust that hath overspread this land,...

...I was a good deal alarm'd at the Bankruptcy of Fordyce[1] and his House as I had just lodg'd a large sum[2] there but was so lucky as to find in the House the day after it stop^{d}, near Two Thousand Pounds worth of Bills which I claim'd and took as my property, they being short wrote in the Account; so that the sum I shall loose by that House will be a trifle but the inconvenience I was at one time apprehensive of from the sudden extinction of Credit throughout this Kingdom was really alarming to a Man like me who have a thousand Mouths to feed weekly wch cannot give Credit; however the storm is over and that without obliging me to be troublesome to my friends.

Fordyce's failure affected particularly the Scottish banks and nearly every private banker in Scotland failed during this period.[3] Boulton weathered the storm as the above letter shows, but to Roebuck, who had sunk his wife's fortune as well as his own in the coal mines at Bo'ness, the restriction of credit was disastrous and he became insolvent. A meeting of his creditors, of whom the firm of Boulton and Fothergill was one, was called, whereat Watt acted as Boulton's attorney, having been so appointed by him in a generous letter dated March 29th, 1773.

Then on May 17th, 1773, Watt gave a discharge to Roebuck for all the sums the latter owed under their partnership. One result of this was that the Kinneil engine became Watt's pro-

[1] Neale, Fordyce and Downe, bankers, failed June 10th, 1772, due to the speculations of James Fordyce.

[2] I.e. a sum of £5000, part of the proceeds of the sale of his wife's Packington estate for £15,000 to Lord Donegall.

[3] Lord, J., *Capital and Steam Power, 1750–1800*, 1923, p. 83.

perty. He had found it to be rusting away, and he therefore had it taken to pieces and the essential parts shipped to Soho via London.

The sum owing by Roebuck to Boulton and Fothergill was some £1200 of which £500 had been advanced on the "faith of being assumed as a partner" in the engine. Boulton hesitated to take over the share in the patent lest the other creditors might think he was receiving an advantage. However, they appear not to have valued it at one farthing so that he was able to get it from the receivers into his hands. Fothergill, however, did not want to have anything to do with the patent or the engine and he was eventually paid his share of the debt of £1200. Afterwards when the engine became a success this led to much acrimony on Fothergill's part although entirely without justification.

Boulton had worked steadily towards his objective, but he had still to withdraw Watt from his surveying business and bring him to Birmingham. The death of Watt's wife on September 24th, 1773, was a great blow to him; he became heartsick of the country and when his last survey—that for a canal between Inverness and Fort William—was finished he shook off the dust of Scotland, like so many of his compatriots before and since, and arrived in Birmingham, May 31st, 1774, to begin a new life. He was given house room by Boulton in the latter's old quarters at Newhall Walk, and we may be sure was made generally comfortable.

Watt's first task was to get the experimental engine to work, and by November he succeeded in doing so as he records in these proud but modest words in a letter to his father: "The fire engine I have invented is now going & answers much better than any other that has yet been made and I expect that the invention will be very beneficial to me".

There was still "a lion in the way", at any rate from Boulton's point of view, and it was this: six of the fourteen years' period for which the patent was granted had already expired and eight years would be obviously too short a period in which to reap the benefit of the outlay that he had foreshadowed in 1769 would be required. There were two courses open: one was to apply for a new patent; the other to get the old one extended.

Watt came up to London in the spring of 1775 and took advice from several friends who were unanimously in favour of the last-named course. Accordingly a petition for leave to bring in a Bill to extend the patent for twenty-five years was drawn up, with Dr Small's assistance, and presented to the House of Commons on February 23rd, 1775.

The course of the petition in Parliament must be outlined briefly. On March 9th the Bill was brought in and read for the first time. On the second reading it met with violent opposition from men like Edmund Burke who were opposed on principle to monopolies of any kind. Boulton came to town in May to help in lobbying for the Bill and the outcome of it was that it passed all its stages and received the Royal Assent on May 22nd, 1775. The Act (15 Geo. III, cap. LXI, p. 1587) extended the patent for twenty-five years from that date and—a point of some importance—extended it also to include Scotland.

While in London Watt received the sad news of the death from ague of Dr Small; this was a great blow to both Boulton and Watt for the successful issue of the patent negotiations had owed much to Small's persistency with the busy Boulton and to his encouragement of the despondent Watt.

During Watt's absence in London, Boulton carried on with the engine experiments as if unto the manner born. An accident happened to the block tin cylinder and a new one had to be obtained; this, in cast iron, was supplied by John Wilkinson[1] of Bersham. The reason for getting the cylinder from him was that in the previous year, 1774, he had invented a mill for boring guns that gave more accurate results than had been obtainable previously. It was equally applicable to boring cylinders. Now Watt's engine cylinders demanded greater

[1] John Wilkinson (1728–1808), the great ironmaster of the eighteenth century, was a man of great force of character but not over scrupulous in his methods. He had a belief in cast iron for every purpose that amounted to fanaticism. He even had his own coffin made in cast iron. We shall hear of him frequently.

accuracy than did the cylinders of the common engine. Hence the invention, coming as it did in the nick of time, was one of the factors that contributed materially to the success of Watt's engine.

Boulton's remaining doubts about the engine now vanished and he became if anything more enamoured of it than the inventor himself. The two men lost no time in entering upon a partnership perhaps the most momentous in industrial history; it began on June 1st, 1775, to run coterminously with the Act of Parliament. The terms of the partnership were briefly that Watt assigned to Boulton two-thirds of the property in the patent while the latter was to pay the expenses already incurred, to defray the cost of experiments, to pay wages, to pay for materials and to keep the books of the firm; Watt was "to make drawings, give directions and make surveys" for which he was to draw £300 a year from the firm. The profits were to be divided in the proportion of their respective shares.

The two men now stood on the threshold of the most important part of their careers, not merely to themselves but to the world at large.

It is not amiss to pause for a moment to review the general situation. It is not realized how ripe was the time in Great Britain for the introduction of some additional source of power. The possibilities of animal, water and wind power for supplying the needs of industry had been explored, as we have already stated, and the only one of these powers that could be exploited further was water and that could only be done by going farther and farther into the remoter parts of the country where water was available. Of all the demands for power those of mine drainage, water supply to towns and feeders to canals, in the order named, were the most insistent. Relief had been afforded by the introduction of the Newcomen or atmospheric engine which had given by this time half a century's yeoman service but its sphere of operations was limited, by reason of its heavy consumption of fuel, to districts such as the coal fields where the cost of fuel was negligible. Further, the engine was not capable of drawing water from a greater depth than about

80 fathoms, so that minerals below that depth had to remain unwon. Power for driving millwork and machinery was required; instead of doing as Boulton had had to do—take the industry to the power—it was the desire to bring the power to the industry and for it to be available to any amount large or small—not limited as at Soho. Such were the problems that the two men faced, hardly conscious of the tremendous import of the issues.

The period reviewed in this chapter with all its troubles was one of enjoyment in the social sphere. Boulton, as will have been concluded already, was of a most hospitable nature; now that he was established at Soho House, he was in a position to indulge his inclination to entertain, not only visitors of distinction of whom we have spoken but also his particular friends. He liked to have them about him and to enjoy their conversation on every topic under the sun. Boulton was no bookworm; rather it was by intercourse with his fellow men that he advanced his knowledge. As Keir says in the memoir from which we have quoted already: "It cannot be doubted that he was indebted for much of his knowledge to the best preceptor, the conversation of eminent men. By most of that description, Mr B. was known and his merit appretiated". Keir goes on to enumerate the many distinguished men of science and of letters that he met at Boulton's table.

It ought to be said that this kind of intercourse was typical of the age and was what gave rise to the salons and coteries which were so marked a feature of the eighteenth century. It was not only in cities like Paris and London that such coteries were formed; they grew up also in the provinces wherever there lived two or three congenial spirits interested in philosophy, science, literature or art. What tended to promote the formation of these coteries was the difficulty of transport which restricted visiting to a radius of a few miles from one's home. It was not feasible then, as it is to-day, to form or become a member of a society with headquarters in the metropolis, to which one can run up to take part in meetings.

Under the aegis of Boulton's hospitality, tact and charming

manners—just such a group, interested more especially in science, grew up in Birmingham. There were like groups in Manchester, Norwich and other places, but none attained to such fame as the first named. The nucleus of the group was undoubtedly Darwin, Small and Boulton himself. Other local celebrities were drawn in: Samuel Garbett, already mentioned; John Baskerville (1706–75), the printer; Samuel Galton (1719–99), the quaker merchant. Singularly enough, John Fothergill is never mentioned as having formed one of the group, nor is John Wilkinson. Then there were others who came at various times to reside in or near Birmingham, like James Watt, James Keir, Richard Lovell Edgeworth, Thomas Day (1748–89), the author, and Dr William Withering (1741–99), the botanist. Others who lived a short distance away came occasionally, such as Josiah Wedgwood from Etruria and John Whitehurst (1713–88), the horologist from Derby. Out of these meetings grew the famous Lunar Society, about which we shall have something to say later. Some of these friends and associates of Boulton are portrayed on Plate I.

This entertaining must have proved costly to Boulton if we are to judge by the amounts of his wine merchant's bills that have been preserved, but he enjoyed it! As one who knew him in these surroundings says:[1] "He was a man to rule society with dignity" and he stood out among his fellows "preeminently as the great *Mecænas*".

[1] *Life of M. A. Schimmelpenninck*, 1858, pp. 32, 34.

CHAPTER V

BOULTON AND WATT

First engine in Midlands—Engines in Cornwall—Engages William Murdock—Keir's sheathing metal—Mechanical paintings—Letter copying—Birmingham Metal Company—Death of Fothergill.

IT is perhaps an exaggeration to say that at the beginning of Boulton and Watt's partnership, the mining and industrial world was gasping for some cheaper and better means of raising water than the atmospheric engine, but there was one part of Great Britain that was in sore need of such means and that was Cornwall. The surface deposits of tin-sand or stream tin had long previously been exhausted and underground mining for tinstone had been going on since the sixteenth century. In the course of this mining, copper ore had been found and was now as important a product as tin. The advent of the atmospheric engine had given relief certainly, but both from the point of view of depth whence it could draw and also from its great cost for coal—all sea borne—mining was approaching a decline.

When rumours arose that a new engine was being introduced, and that it was under the aegis of such a responsible man as Boulton, the Cornish adventurers began to make enquiries about it. Boulton replied judiciously to his correspondents and raised their expectations. However the adventurers were unbelievers and cautious to a degree, so they preferred to sit on the fence and await events.

It was now Watt's turn by performance to implement Boulton's promises. So far only the results of the experiments on the Kinneil engine were available and Watt, we may be sure, would have liked to have made haste slowly by building another engine, not too much bigger, in order to gain experience. Not so the impetuous Boulton, who wanted to produce a big engine that would serve for show purposes. Boulton was a great believer in advertisement—Soho Manufactory was a standing example. Watt went to the other extreme; he would have preferred to have hidden his light under a bushel. Boulton carried the day and took an order for an engine at Bloomfield Colliery, Tipton, Staffordshire, not far away, with the proprietors of which, Messrs Bentley and Company, he was acquainted. The cylinder was 50 in. diam. which was a big jump from the 18 in. cylinder of the Kinneil engine. Bloomfield engine was set to work on March 8th, 1776, with much fanfare of trumpets; a long account appeared in the local paper,[1] the first public mention of it of any kind in print; it was a great relief to Watt when it proved, as it did, a success.

Practically at the same time a blowing engine, smaller in diameter—38 in.—was made for John Wilkinson, who always insisted on having the very latest improvement, to blow his blast furnace at Broseley Ironworks, Shropshire. While Watt was there, Boulton wrote to him (1776, Feb. 24): "I have fixed my mind on making from twelve to fifteen reciprocating and fifty rotative engines per annum. I assure you that of all the toys we manufacture at Soho, none shall take the place of fire engines in respect of my attention." By "rotative" Boulton meant the rotary engine or steam wheel that Watt had invented but that did not survive.

Broseley engine was likewise a success; judged from a mechanical point of view, both it and Bloomfield engine were badly made and consequently costly in upkeep. However, they served the intended purpose—that of launching the engine on the world, convincing the doubters and permitting the firm to gain experience. What the engine was like at this period may be judged by the illustration (Fig. 3). The original, drawn by Watt himself, is entitled "General Section of Engine" which suggests that he considered it representative. Actually it was the engine for Torryburn, Fife, designed in 1776 and erected in 1778.

The method of building these and succeeding engines was not that of *manufacturing* foreshadowed by Boulton in 1769 (see p. 81). The partners took the line of least resistance and followed the practice of the builders of the common engine, that

[1] *Aris's Birmingham Gazette*, March 11th, 1776.

is to say the proprietor or client assembled the materials from different parts of the country and paid for them; they were put

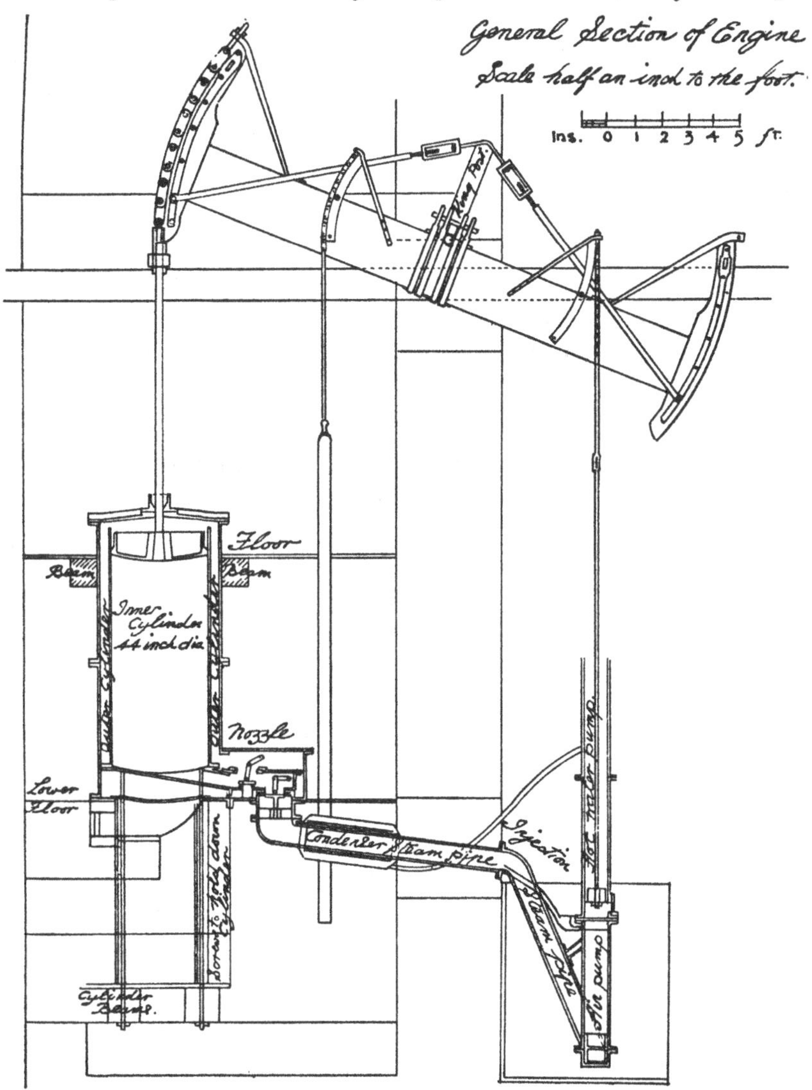

Fig. 3. General Section of Watt's engine, 1776.
Boulton and Watt Collection. Courtesy of the Public Libraries Committee, Birmingham.

together on the spot by his own craftsmen to drawings and instructions furnished by Watt. A few special parts only, like the "nozzles" or valves, were supplied from Soho. There was one exception to this rule: the firm did insist on the cylinder—the vital part of the engine—being supplied by Wilkinson for the reason already given—he alone could bore cylinders with the accuracy required. Writing on this point to Garbett[1] at Carron, who thought the Smeaton boring mill there was quite capable of turning out a cylinder for an engine he was proposing to build, Boulton said: "Wilkinson hath bored us several cylinders almost without Error; that of 50 Inches diam^{r} for Bentley & Co. doth not err the thickness of an old shilling in no part, so that you must improve in boring or we must furnish the cylinder"; and again (February 17th) Boulton writes: "I am a great lover of truth both moral and Geometrical—Bore your cylinders as true as Wilkinson's and then say there is no truth in me if we are not good customers to Carron." The great superiority of Wilkinson's boring mill lay in the fact that it turned out a cylinder that was not only circular at any section but was truly cylindrical throughout its length. For twenty years and more Wilkinson retained this monopoly from Boulton and Watt.

There was one thing further that the firm stipulated with their clients and that was that the erection was to be in charge of one of the firm's own men whose wages and expenses were, however, to be paid by the client. A new class of men of superior intelligence and technical ability thus came to be required as erectors; for many years such men could only be had by training. Their many lapses from time to time were a constant source of irritation to Watt; Boulton with his wider knowledge of mankind was ever tolerant of their shortcomings. It will be realized that by adopting the old system of engine erection, little space at the Manufactory beyond a yard at the rear needed to be allotted to the new business. Further Watt made his calculations and drawings and carried on the correspondence at his own residence at Harper's Hill. By this arrangement it will be seen that the partners acted much more as consulting engineers than as engine makers. Hence it will be asked how they remunerated themselves for their services, and the reply is that the terms were

[1] B. and W. Coll. 1776, Feb. 10.

the value in money of one-third of the savings in fuel of their engine as compared with a common engine.

Having got the Bloomfield engine off his mind, Watt went back in June to Scotland for the important purpose of taking unto himself a second wife. His prospective father-in-law James MacGregor, a dyer in Glasgow, cautious Scot that he was, before parting with his daughter, wanted to see the deed of partnership with Boulton. Astonishing as it may seem, the cautious Watt and the businesslike Boulton had not even then executed the deed, such was their confidence in one another. Boulton got over Watt's predicament by writing an ostensible letter, which can only be described as a prevarication (1776, July 3):

> I would without hesitation have sent you the assignment and the article of partnership had it been in my power, but Mr. Dadley, the lawyer, is suddenly called to London, and it cannot be had before his return; but if you want to show it to any of your friends you may give them a copy of the several heads which I have extracted from our mutual missives, and are to the best of my knowledge all that our articles contain.

Then follow the articles of which we have already given a summary. Among all the papers of the partners this deed is not to be found and we are left to surmise whether the deed was ever executed or not!

Reports upon the performances of the new engine were spread abroad and other engines were ordered farther afield. The first one in the London area was one for Cook, Adams, Wilbie and Sager, for their distillery at Stratford-le-Bow. It was 18 in. diam. and was started in December, 1776. In the following year the cylinder was found to be badly worn, due, as Watt reported, to the fact that "the devil prompted them to put sand in the piston, in which way it wrought a long time", a woeful example of misplaced ingenuity! It was this engine that John Smeaton, the celebrated engineer, visited in 1777, with the sad results as reported by Boulton (1777, April 20):

> Hadley [i.e. the firm's erector]...told me that Mr. Smeaton & Holmes called upon him to request him to attend them to Stratford-le-

Bow, which he did. Smeaton said it was a pretty engine, but it appeared to him to be too complex; but that might in some degree be owing to his not clearly understanding all ye parts. He gave the engineer mony to drink & the consequences of that was that ye next day the engine was almost broke to pieces. Wilbey was very angry turned away the engineer, and told Hadley the least amends he could make was to put it in order again wch he did do, but was obliged to put in one new valve.

Smeaton's opinion could hardly have been otherwise than one-sided. He was prejudiced in favour of the common engine because he had spent much time on its improvement and had brought it to the highest pitch of excellence of which it was capable. Later he completely changed his opinion about Watt's engine.

Space does not admit of treating the further progress of the engine in the same detail as the foregoing; we can only pause to point out the more important stages. One of these was the introduction of the engine into Cornwall. A deputation of adventurers had been in the middle of 1776 to see the Bloomfield engine but were still dubious about its performance, and negotiations went on so that it was not till 1777 that the firm received an order for one 52 in. diam. for Tingtang Mine near St Day, followed quickly by another of 30 in. diam. for Wheal[1] Busy, near Chacewater. To these Watt gave special attention as upon them would depend further orders from Cornwall. The materials were ordered in May and in August Watt accompanied by his wife went down to Cornwall to view the spot. He spent four days *en route*, such were the travel conditions of the period. We cannot go into the difficulties and perplexities that had to be surmounted. Suffice it to say that, although Watt would have liked to have started the bigger engine first, Wheal Busy was ready first, and began working in September, while Tingtang was not ready till January, 1778.

As soon as Wheal Busy engine had shown her paces, two larger engines, both 63 in. cylinders, one for Tregurtha Downs, a mine of the Wheal Union group and the other for Chacewater,

[1] "Wheal" is the anglicised form of Cornish "Huel", meaning "mine".

were ordered. The former was set to work in August and the latter in September, 1778. Of the former Watt wrote (1778, August 13): "The conoissieurs say we shall never fork ye water or *if* we do we may fork anything as ye water is reckoned the heaviest in ye whole county." The "conoissieurs" were confounded and the result was a flood of orders. By 1780 the firm had erected forty engines, twenty of which were in Cornwall. In 1783 Watt informed his partner that there was only one common engine left in Cornwall.

It has been stated above that the engines were paid for on a royalty basis. This was a scheme of Watt and some idea of its nature and how the sum payable was calculated, must be given. This is explained in a letter from Watt to Jonathan Hornblower, the best known engineer of that time in Cornwall (1776, October 17):

> Our profitts arise not from making the engines but from a certain proportion of the savings in fuel which we make over any common engine that raises the same quantity of water to the same height. The proportion of savings we ask is one-third part to be paid to us annually for twenty-five years, or if our employers chuse it, they may purchase up our part at ten years price in ready money.

There were very few cases where a simple replacement of a common engine by a Watt engine took place and actual measurements could be made, because usually the new engine had to pump more water or it had to pump from a greater depth than the old one; then too the mine for which it was wanted might be an entirely new one. To establish a standard of comparison, two common engines at Poldice Mine were agreed upon as average and a small committee made the necessary tests. It found that the "duty", which is the Cornish way of expressing the performance of an engine and meant the number of pounds of water raised one foot high per bushel (say 94 lb.) of coal, was seven million. The load on the piston of the common engine when performing its best was 7 lb. per sq. in., while Watt's engine did its best with a load of 10½ lb. In other words, to do the same work the Watt engine was smaller than the common engine in the ratio of 3 : 2. Watt drew up a table of sizes of his engine with the corresponding sizes of common engines and the appropriate figure was inserted in the agreement that was entered into. To calculate the savings, it only remained to measure the coal consumed and the quantity of water raised. The former was already being done on the Cornish mines, because by an Act of Parliament of 1751, a drawback of the duty on exported coal was allowed on all coal consumed in pumping there. To measure the water pumped, knowing the diameter of the pump barrel and the stroke, a figure could be calculated giving the weight of water in pounds delivered at every stroke. It was finally necessary to count the number of strokes and this was done by a mechanical counter fixed to the beam of the engine in a locked box to prevent it being tampered with. Boulton got the idea of this counter in 1777 from a pedometer, made by the firm of Wyke and Green in Liverpool, and made some of them himself at Soho.

This method of charging by royalty or premium as Watt terms it, was quite fair because the adventurers only paid so long as the engine was working. If the mine closed down, the payment stopped. However, the method of calculating the premium was not too readily grasped and, much to Watt's chagrin, although probably to Boulton's secret satisfaction, the partners had to give way in favour of a fixed payment annually for each engine according to its size.

Watt returned to Birmingham in September, 1777, having seen the engines of Tingtang and Wheal Busy going satisfactorily and the reputation of Boulton and Watt established. In May of the following year, he had to go again to Cornwall for the firm and we must leave Watt to face his troubles with other constructional details such as piston packing and his no lesser difficulties with the mine adventurers with whom he was constitutionally unable to cope; in despair he had to invoke his partner's aid. Boulton managed to pay his expected visit to Cornwall in October; with his knowledge of men and skill in business negotiation, he was able to arrange the agreements not only for the engines in prospect but for those actually at work. The amounts payable in premiums were now quite handsome and to a man like Watt, accustomed to poor circumstances, seemed beyond the dreams of avarice. To Boulton, the prospect was very different. He was in need of fresh capital to finance his greatly increased business other than the engine one, and he visualised in these premiums a form of security that he could offer to bankers to get them to advance the money he wanted. It is not easy for us in these days of joint-stock enterprise, when capital is obtained so readily, to realize that in the eighteenth century it was far otherwise. Bankers were all private individuals and, as at present, would only grant temporary accommodation. Short period loans could be obtained by bills and mortgages. Of course it was possible to take a partner with capital and this was the usual course, but one had to be circumspect. As the law stood, liability was unlimited and the failure of one partner could reduce the others to beggary. Now Boulton had a number of partners—a list of these at different times is given on p. 209—and a suggestion to take another multiplied risk instead, as one would have thought, of reducing it.

Boulton was successful in obtaining £2000 from Elliott and Praed, bankers, of Truro, and he obtained also a large sum from Lowe, Vere and Company, London bankers. Watt's consent to pledging the securities had to be obtained and this burden of debt made him utterly despondent and wretched, quite without real reason, for the Manufactory was working at full pressure and the engine business now showed that it was going to be a success.

The Cornish engine business developed to such an extent that it necessitated the presence of one or other of the partners in the county practically the whole time. They had appointed an agent there who acted for them in money matters, Thomas Wilson, of Truro, who served them faithfully; they had, of course, their erectors on each of the mines where they were erecting engines, but in general these men were irresponsible and easily overcome by drink.

A great exception to them was William Murdock, who came to Soho in 1777 to look for a job. Boulton interviewed him and observing that the hat, which the bashful young man was nervously twiddling, looked peculiar, elicited from him that it was of wood and had been made on a lathe of his own making. Only an oval lathe would do such a job and this was evidence of unusual knowledge and mechanical skill. It was enough! Murdock was engaged as an engine erector and, after proving his value on various jobs near home, was sent to Cornwall in 1779; there he quickly came to the front, for, besides being intelligent, resourceful and inventive, he was able to resist strong drink, indulgence in which was the great failing of those days. He lived to become the right hand man of the firm in Cornwall—a tower of strength true to the firm's interest—and remained in the county till the expiration of the patent approached; resisting all offers of advancement there, he returned to Soho, where he was equally valuable, and remained with the firm till the day of his death.

The fame of Watt's engine spread abroad and one of the earliest to get to know of it was J. C. Perrier (1742–1818), a smith and a clever mechanician, who was in partnership with his brother, A. C. Perrier. The former was a pushful man and he obtained an *arrêt de conseil* or decree authorizing him to raise water from the Seine to supply Paris with water. Unable to make the engine, he came to England and visited Wilkinson at Bilston. Watt learnt from the latter in May, 1777, that Perrier's original intention was to construct an atmospheric engine but that, hearing about Watt's invention, he wished to install one of his improved type. He wanted Wilkinson to make it but the latter declined because he wished to remain on good terms with Boulton and Watt; otherwise he would have been liable to an action for infringement. This put the partners on their guard. Meanwhile, the Count d'Heronville had approached the firm for an engine to drain some marshland near Dunkirk. The firm consented on condition that the Count would get an *arrêt de conseil* for them. This he succeeded in doing, the condition being that the engine on trial should show its superiority over the common engine. Nothing, however, came of the Dunkirk scheme but a certain M. Jary, concessionaire of coal mines near Nantes, asked the firm for an engine for draining his colliery.

He, too, obtained a decree with slight modifications. Thereupon Jary was supplied with drawings, instructions and the usual nozzles, the latter being exported with some little difficulty owing to the state of warfare between the two countries. Knowing possibly that he could not be brought to book, Jary never paid a penny piece for his engine. As we do not hear any more of it, perhaps he could not get it to work—if so, it was a just retribution.

Perrier came again in January, 1778, and had perforce to come to the proper quarter. However, the partners decided to give him easy terms for the decree had no effect in law till its conditions were fulfilled and that could only be after the erection and setting to work of the engine. Perrier eventually erected two engines at Chaillot-sur-Seine, then outside but now within the boundary of Paris. There they worked for many a long year, one of the wonders of that city.

Seeing that one or other of the partners was obliged to spend so much time in Cornwall, Boulton took and furnished a house at Cosgarne near the mining area but secluded from it, and there they lived with as much content as they could summon.

The absences of the partners in Cornwall, while disagreeable to Watt in that he was divorced for the time being from his comfortable home and from the society of his friends, were not only equally disagreeable to Boulton, but interfered seriously with the various other businesses in which he was engaged. From this time onward we observe a decline in the attention he devoted to artistic productions. To a certain extent this was the outcome of that trait—common among creative persons and eminently observable in Boulton—that a man loses interest in a job after he has thoroughly mastered it and sighs for fresh fields to conquer, but principally the decline took place because these artistic branches did not pay.

It was a pressing need for Boulton to have someone to whom he could delegate his authority when absent. In those days it was usual for men in a large way of business to have a clerk who acted as manager. With such an extended field of operations as Boulton had, he needed someone with higher qualifications than a clerk, and considered seriously the introduction of another partner. Through Darwin he had got to know James Keir (1735–1820), who has been mentioned already several times.[1] He was a man who had had a varied career. Darwin and he had been schoolfellows together and while both took up medicine at Edinburgh University, Keir threw up his studies in order to enter the army to satisfy his desire of seeing foreign countries. He had his wish and attained the rank of captain. His marriage about 1770 to a Miss Harvey led him to resign from the army and his friendship with Darwin brought him into the Midlands. He took up again his College studies in chemistry and being of an original mind, he entered into the business of manufacturing glass at Stourbridge about 1775. Darwin introduced Keir both to Wedgwood and to Boulton. We find the latter in 1777 sending Keir a silver coffee pot and lamp as a present to Mrs Keir.

What more natural than that Boulton should think of Keir to assist in the management and take charge of Soho during his absence! Incidentally it was not to be expected that Watt could undertake this task, for not only did intercourse with strangers give him violent headaches, but he was not on the spot for, as we have said, he worked at his house at Harper's Hill, nearly a mile distant from the Manufactory.

The story of Keir's connection with Soho is best told in his own words in the memoir on Boulton to which we have referred already:

I resided at Stourbridge and was engaged in a glass manufactory in partnership with old John Taylor, Skey of Bewdley, &c. I was the managing partner, for which I had a sellary, and a share of the profits, besides my share proportionate to the capital I had in the business. Mr B. with whom I was in intimate friendship, finding that the management of his manufactory and other great concerns, was too much for any one man, expressed a wish that I should join him in partnership and management of his manufactory. I accepted the proposal but thought it necessary that Mr Fothergill should join in the invitation to me. Accordingly they sent me a letter signed by both to that effect, & offering $\frac{1}{4}$ of the profits for management & a

[1] Moilliet, J. Keir, *Sketch of the Life of James Keir, Esq., F.R.S.* [1868].

proportionate profit for any capital I might hereafter add to theirs. Upon this I gave up my establishment at Stourbridge & removed to my house at Winson Green. I then visited Mr B. & Soho daily. We examined together the details of some of the business & saw that considerable amendments in the economical part were requisite. Mr B. consulted me on many subjects relative to his business & to his correspondences. I perceived that there were great financial embarrassments. Mr Fothergill told me the amount of their discount or bill account which appeared quite alarming. Mr Walker confirmed it for I never looked into their books of accounts. It appeared evident that the business had been carried on with great loss, as it was afterwards proved to have been, by the insolvency of Mr Fothergill at his death.—Under these circumstances, the intended partnership could not with any propriety be compleated, & it was never mentioned by either of us. Mr B. continued to consult me about his affairs which he had the most sanguine hopes of re-establishing by the profits he expected from the fire-engine business. After waiting two years, without receiving any intimation of any plan for a connexion, I gave up all expectation, supposing it to be impracticable, and besides could no longer afford to be idle having already suffered considerably in my small fortune from having relinquished a tolerable establishment at Stourbridge, without having any means of compensating for the loss, not being entitled to nor having received any emolument from Soho, excepting in one instance, namely the silk reels, of the profit from which, Mr B. upon his return from Cornwall, ordered ¼th (which was the proportion intended if the partnership had been completed) to be put to my credit, tho' I had made no stipulation or demand. I do not remember the amount, but I think it was near £200. [It was actually that amount.] I can make no doubt that Mr B. wished he had had it in his power to have made a different result, but I believe his difficulties & restraints were insurmountable. I confess that I was vexed that he was not sufficiently explicit to tell me but that soon passed, and a very sincere & hearty affection for him continued to the last.

In spite of the fact that Boulton offered Keir no partnership or definite position in the works, he was not idle. In 1779 the latter took out a patent 10th Dec. (No. 1240) for a "compound metal capable of being forged hot or cold, for making bolts, nails, sheathing for ships and for other purposes". The alloy consisted of 100 parts of copper 75 of zinc or spelter and 10 of iron.

But before the patent had been granted, Boulton had with his usual energy taken up the exploitation of the alloy. On August 14th, 1779, he addressed a charmingly polite letter[1] to the Earl of Sandwich (a Montagu and, therefore, a distant relation), first lord of the Admiralty, asking to be allowed to wait upon his lordship respecting a memorandum which Boulton submitted with his letter. The memorandum is worth quoting:

Matthew Boulton and James Keir have discovered a compound metal which they apprehend would be considerably better than copper for sheathing ships, and for making the bolts and nails and other pieces which are now made of copper. Their reasons are:

1. That the metal being much harder and stronger than copper or any metal except forged iron, to which it is very similar in those properties as also in its being sufficiently ductile and tough, is much fitter for the purpose of making bolts and nails as they may be drove better and further into the timbers, without being upset, as the carpenters call it; and when drove they are much less liable to bend, and consequently give great strength to the ship.

2. That the metal is less liable to be corroded or dissolved by salt water than copper is, and consequently would last longer.

3. That they engage it shall not cost more than copper—perhaps less.

4. That they request leave to have a trial made, and if their request be granted they will furnish a sufficient number of bolts, nails and plates for that purpose.

The matter was referred to the Navy Board and an experiment was ordered to be tried on the "Juno" frigate, as appears from the following letter:

London October 8th, 1779.

Honble Sirs—In consequence of the permission which you were pleased to give that the bolts and plates made of a new metal invented by Mr Boulton and myself should be tried on the Juno frigate now building in Batson's yard, I have prepared a sufficient quantity of those bolts and plates for such trial, and shall be ready to attend at Batson's yard for that purpose on Monday next or any day after that which you shall be pleased to appoint.

I am &c. James Keir.

P.S. I wait to receive the honour of your commands.

[1] Prosser, R. B., *Birmingham Inventors*, p. 118.

He must have taken the letter in person to the Navy Board, as the letter is endorsed the same day: "Mr Keir was desired to attend the Board on Tuesday next between ten and eleven".

A trial of the bolts took place at Deptford Yard and a report upon them—not favourable for they were too soft and so "upset" in driving home—was made to Sir John Williams and Edward Hunt, Surveyors to the Navy. Thus damned with faint praise, nothing further came of the metal as far as use in the Navy was concerned.

On June 11th, 1781, Boulton and Keir executed a mutual discharge from all claims and demands on one another. But that was not the end of the alloy for before that date, i.e. in 1780, Keir left Soho to enter into partnership with Alexander Blair to establish a works at Tipton, Staffordshire, to make window sashes, etc., from the alloy; incidentally this proves that metal casements are a good deal older than is usually supposed. Shaw in his *Staffordshire* gives an account of the works and says of the alloy that it was "a peculiar golden coloured metal". Window casements were supplied to Windsor Castle, Carlton House and "many principal mansions in the Kingdom" including, we believe, Soho House.

The author has not been able to find why, unless it was on account of cost, these casements did not continue to be made and why they fell into disuse. The similarity in composition of Keir's alloy to Collins's metal and to the well-known and successful Muntz's metal, was pointed out in 1866 by Mr W. C. Aitken and is most readily shown by giving their percentage compositions side by side, thus:

	Keir, 1779	Collins, 1800	Muntz, 1832
Copper	54·05	55·5	60
Zinc	40·55	44·5	40
Iron	5·40	—	—

All that we can say is that Keir was before his time.

Keir's memorandum shows that he was disappointed that the mooted partnership with Boulton and Fothergill had proved impossible and that consequently he had wasted opportunity and time. Boulton must have realized it too and it may have been a sop to Keir that he was taken into the copying press business (see p. 108).

We may perhaps mention the subsequent career of Keir. He and Blair at Tipton made alkali for soap making by a process discovered by the former. In 1794 these partners bought land at Tividale, Staffordshire and sank pits there known as the Tividale Colliery—a successful venture.

We must now leave the steam engine for a while, but not the Manufactory, in order to give an account, as clear as is possible with the scanty material that is available, of how Boulton engaged in yet another branch of artistic production—that of copying pictures—and this in spite of the fact that he had been bitten over ormolu manufacture and that his hands were full with the steam engine. Any proposition of an art nature seems to have had a peculiar fascination for him; however, we cannot but believe that in undertaking picture copying he was acting in opposition to the better judgment of his partners.

It is fairly certain that the idea of copying pictures was not his own but that it emanated from Francis Eginton (1737–1805), an artist living in Birmingham, who had been employed at Soho as a decorator of japanned ware. What we find is that he was taken into partnership or perhaps only engaged by Boulton for the production of "mechanical paintings". Correspondence shows that this work began as early as June, 1778; what the process employed was, research has quite failed to reveal. We have only the most meagre scraps of information and all that we can be sure of is that the process was some modification of the engraver's art with subsequent additions by hand. Boulton gives a hint as to how he went about it in a letter (1779, June 12th) to Sir Watkin Williams Wynn, 4th baronet, asking for the loan of pictures. Boulton states: "I am engaged in painting as a manufacture and that by some peculiar contrivances, I am enabled to make better copies of good originals than can be done otherwise, without much greater expense; or rather by multiplying those copies when once obtained and making an extensive sale of them in foreign countries".

We get a crumb or two of information from sentences in Eginton's letters such as: "Take off a few impressions for me" and "There are two plates for one picture and three for another". Joseph Barney (1751–1827), an engraver of Wolverhampton, who was employed in turning out these pictures and seems to have done a great many, speaks of "the printed impression". The suggestion is that the impressions were made with a copper plate printing press and that there were two or more plates applying different colours. Quite a number of pictures are mentioned as having been reproduced. That there was a large amount of actual painting in colour required, appears, for instance, from the fact that in 1781, May 13th, Barney sends in a bill for paintings, items in which are 15, 8 and 5 guineas. On August 29th, he says: "Each picture will take me 12 days of 10 hours and if I paint them at 7 guineas each you can judge the profit I shall make by them".

As to the method of painting, Barney on May 17th says: "I regret Mr Boulton is not pleased with the last pictures. If he will return the printed impression of the Stratonice, I will endeavour to make as good a picture of it as I can". Then on September 20th he says: "The last picture...cannot be finished without being dead coloured again".

As to the materials used we have this in a letter from Eginton (1782, March 8th): "I have no objection to keeping all the 24 yards of canvas, having orders for pictures that will nearly work it all up". The work went on as late as 1791 for, on April 15th, Eginton sent in a bill for £12. *6s.* *6d.*, his charge for "finishing from the dead colour and retouching 4 pictures for Mr Boulton".

The manufacture of these paintings was never remunerative and Fothergill was insistent that it should be stopped. In 1779 he returned the loss on this business as £423. 12*s.* 5*d.* and again writing to Boulton (1780, February 1st) he said: the "Painting business has turned out £500 on the wrong side. Judge if such a detrimental business should any longer be pursued". Consequently in June, 1780, the agreement with Eginton was terminated, but work still went on with Barney as is seen above. However 1781 seems to have seen the end of it, for these replies are sent to prospective customers: "We now only take orders for such subjects as we have apparatus for" and "Mr Boulton's time is wholly engrossed by his steam engine business. He has lately resolved to entirely drop painting".

It is of interest to know that, in 1784, Eginton set up as a glass painter not far from Soho, and achieved notable success. Many stained glass windows, e.g. in St Philip's Church, Birmingham, remain to-day to testify to his talents in this direction.

It only remains to say that there is a curious parallel with Eginton and Boulton's work in the production of mechanical pictures; we hesitate to call it a sequel, but it is this:

In 1784 an artist named Joseph Booth, of Golden Square, London, brought out a method of "multiplying pictures in Oil Colours" without the aid of an engraver; to the process he gave the formidable name of "pollaplasiasmos". He exhibited a specimen and issued a pamphlet to announce the exhibition. In a further pamphlet written, from internal evidence, between 1784 and 1786 entitled: *An address to the Public, on the Polygraphic art, or the Copying or Multiplying Pictures in Oil Colours, by a Chymical and Mechanical Process*, we observe that the name has been changed and we learn that the process is a chemical and mechanical one, but he carefully refrains from giving any explanation of it; in the text he mentions "the experience of twelve years" so that this would throw back the date of the original invention to, say, 1774, i.e. anterior to the date of the beginning of Eginton's process. We need only say further that Booth founded the "Polygraphic Society" which held annual exhibitions from 1784 to 1794. Large numbers of pictures were produced by the process and some still exist; these pictures have been mistaken for genuine oil paintings, even by experts. The Society came to an end in 1794.[1]

[1] For a full account of the invention with a technical explanation of the process, based on the known facts and examination of actual polygraphs by a practical colour printer, see H. G. Clarke, *Printing Review*, vol. III, 1933, pp. 61, 131, 175.

Let it be said at once that there is no mention of Booth in the Boulton papers. The only hint that we have found that there was any connection between the process of Eginton and that of Booth is contained in the following remark by Keir in the memoir of 1819 already so extensively drawn upon:

> The paintings which were called mechanical (now styled polygraphic) which might be called *painted impressions,* had not an extensive sale, proportionate to their execution. The more a work of taste is multiplied, so that many may possess it, the more its imaginary value is diminished.

Watt, too, in his memoir, seems to connect Eginton's work with that of Booth by calling the former's productions "what are now called polygraphic pictures".

However this is not conclusive and all that it is safe to say is that both artists were working on the same problem, as many had done before and many have since. Hence, until evidence to the contrary is produced we conclude that Eginton's "mechanical paintings" were not the same as Booth's "polygraphs".

We have seen that Eginton began his process of copying pictures in 1778. Is it too much to suppose that Watt, cognisant, of course, of what was going on at Soho, might have thereby had his mind turned in a similar direction? Ordinarily it was only necessary to give him a hint about an invention and he would pounce upon it and develop it further than the inventor himself had done. However that may have been, in 1779 Watt *did* invent an analogous copying process except that it was for correspondence and documents—subsequently known as press-copying. To tell the story requires some little digression. His absences from headquarters, especially in Cornwall for extended periods, threw upon him a load of correspondence and drawing of which he was bound to keep records. This could only be done by slavishly making copies of everything he did. He had no one to help him, and naturally he found the task most irksome. Stimulated by the desire to reduce the drudgery, he invented writing with ink mixed with gum arabic or sugar and pressing the writing against an unsized damped sheet of tissue paper, with the result that the writing is offset and can be read on the reverse side, i.e. by looking at the back of the sheet.

Boulton was the first to whom Watt communicated the invention in a letter (1779, June 28) from Cornwall. He proceeded to complete the invention by designing a roller press to make the impressions. Here the analogy with the copper-plate printer's press is obvious. He envisaged also the screw-down press which at a later date, when press-copying came into use generally, was found more convenient and was therefore preferred. Perhaps we ought to inform the present generation, ignorant of aught else but the carbon copy made by the typewriter, that press-copying was in use universally for well over a century; it is difficult to imagine how business could have been carried on during that time without its aid.

It is more than likely that Watt would have been content to have kept the invention for his own and the firm's use, but Boulton, with his usual acumen, saw the wide possibilities of the invention and decided to take it up commercially. Letters Patent (1780, Feb. 14, No. 1244) were taken out by Watt and in the specification the screw-down press as well as the roller press is described. On March 20th, 1780, a partnership of Watt, Boulton and Keir (the latter perhaps brought in, as we have said, because of his disappointment with regard to a partnership with Boulton and Fothergill) was formed to manufacture the apparatus. Boulton was enthusiastic; he took a press to London and showed it to the Members of both Houses of Parliament, to bankers and to merchants in the city. As might be expected, the invention encountered opposition at first. Bankers especially looked upon it with disfavour from the belief, unfounded of course, that it would encourage forgery! Nevertheless there were many who could appreciate a good thing and 150 presses were made and sold in the first year; a steady business was carried on during the fourteen years' duration of the patent.

The press was made of foolscap size but a still larger size for copying drawings was made and used extensively by the firm; as the paper for this was cartridge or of that thickness, the copy appeared to reverse hand so that lettering and figures had to be inserted by hand on the copy after printing. Hundreds of these drawings marked "reverse" are to be found in the Boulton and Watt Collection.

When the patent expired, the business was still carried on, new life being infused into it by James Watt, junior, as will appear on a subsequent page.

Boulton and Fothergill were vitally interested in the supply of raw materials for their Manufactory and one of these was brass. At this time Birmingham was dependent upon the manufacturers at Cheadle, Staffordshire, and upon the importers of Bristol for the supply of this alloy. These two bodies, acting in concert, in 1780 put up the price of brass £13 per ton and this led to a proposal by the brass founders in Birmingham that they should make their own brass. The initiative in this matter was taken by Peter Capper of Bristol, who tried to draw in Boulton, but he was not particularly attracted by the proposal. A meeting was called for November 28th, and Boulton submitted a scheme for a "Birmingham Metal Company", which was adopted. Subscriptions were invited in the following February and the capital was over-subscribed. Friction arose over the site to be adopted for the brass works. Boulton and Capper were in favour of Swansea but were over-ruled and a site in Birmingham was chosen. This led to Boulton's, and later to Capper's, resignation from the Committee. Boulton's vanity seems to have been hurt by the rejection of his advice and he prophesied failure. His prophecies were belied by the success of the Company over many years, success which was ensured by the fact that its members were consumers who furnished an assured market for its product.[1]

The partnership of Boulton and Fothergill was an unsuccessful one. From a statement prepared by Zacchaeus Walker, the clerk to the firm, it appears that on a capital of £20,000, the excess of losses over profits for the eighteen years of the partnership ending 1780 was upwards of £11,000. Had it not been for the money Boulton received from the sale of his wives' estates, the firm must inevitably have gone bankrupt,

[1] Roll, E., *An Early Experiment in Industrial Organization*, 1930, pp. 95–97.

because the profits from the steam engine business had not yet materialized. Why the hardware business should have been unsuccessful is not clear, but there were faults on both sides. Fothergill seems to have been of almost as despondent a character as Watt. Boulton must many a time have been hard put to it to keep up his spirits with two such partners, one of them grousing in Birmingham and the other groaning in Cornwall. Fothergill had established many foreign connections but they did not prove profitable to the firm, and Mathews, the London agent, had urged that this foreign business should be closed down. Boulton, we must admit, was far too ready to take up new branches of manufacture; he already had too many irons in the fire and consequently he was unable with all his energy to devote adequate time to each. Even in 1774, Fothergill considered the firm to be on the brink of ruin on several occasions; in despair he suggested going into bankruptcy: "Better stop payment at once, call our creditors together and face the worst than go on in this neck-and-neck race with ruin".

Boulton, however, would not hear of it; probably his pride was involved; besides there was the moral certainty that if one business smashed so would his other businesses—there was no such thing as limited liability in those days. Fothergill was possibly induced to carry on in the hope, faint perhaps, that the corner would eventually be turned and that if he were to be ruined, he might as well be hung for a sheep as for a lamb. But the situation could not last. It culminated when Fothergill, so it appears, made charges that Boulton had cheated him. The latter responded by giving his partner notice (1781, Nov. 2) that the partnership should cease on December 31st of that year and apparently the notice took effect.

In the course of a letter to Watt (B. and W. Coll. 1781, Nov. 13), Boulton gives particulars of a long conference on the subject and of his decision to allow Fothergill or his family something out of the prospective engine profits: "I have suffered my pity to predominate over my Judgment & have given him or intend to give him rather more than is consistent with my

Stewardship to my own Family: however, peace and happiness cannot be bought too dear".

Fothergill became insolvent and, overwhelmed by his misfortunes, died in 1782. A dispute over money matters arose, the details of which it is not easy to follow, but it was submitted to Samuel Garbett and Samuel Galton as arbitrators; they found in an award dated July 29th, 1783, that a sum of £19,284 odd was due to Boulton from the estate of the late partnership of Boulton and Fothergill. What Boulton got from this award, we suppose, could not be money, but he got Soho Manufactory with all its appurtenances and goodwill, so that henceforth it became his undivided property.

It will be recalled that Fothergill had borrowed from his friends the money with which to start business. He had been unable to repay any of it, indeed he had got deeper into the mire. In February, 1773, in a letter to Boulton he said: "I have already assisted the business with upwards of £6000 of Mrs Swellingrebel's money without giving the least security". This lady was the widow of a Dutch merchant in Amsterdam, from whom possibly Fothergill had borrowed the original sum he started with. Whether this sum of £6000 was part of the original sum or a fresh loan, we do not know, but as he was now unable to pay the interest, she was almost destitute. Fothergill left a wife and seven children who were in like plight. Boulton acted with the greatest consideration towards both families. He paid an annuity to Mrs Swellingrebel and out of the engine profits helped Mrs Fothergill with the education of her children. In his will, as will be seen later, he left each of the sons a small legacy.

It is not to be inferred that the dissolution of the partnership with Fothergill disturbed in any way the activities of the Manufactory. It still went on but we imagine that a resolute retrenchment in and overhaul of the business took place to put it on a paying basis. Perhaps it was for this purpose that Boulton, to replace Fothergill, engaged John Scale. There may have been a partnership between them, but it is a peculiar fact, so far unexplained, that these so-called partnerships do not seem to have been implemented by partnership deeds, or at any rate they are not now to be found. This is the case with Watt, the partnership with whom was surely one that ought to have been legally secured, but, as we have seen in a previous chapter (p. 93), it does not seem to have been done.

Possibly also the fact that the premium payments from Cornwall were now coming in eased the situation. Up to December 31st, 1780, £2600 had been received and Boulton was now paying off the loans and overdrafts. It was not till about 1785, however, that he had freed himself from his entanglements.

A great deal happened, however, before that and to deal with it requires a fresh chapter.

CHAPTER VI

THE ROTATIVE STEAM ENGINE

Watt's patents—Boulton's inventiveness—Death of his second wife—Breakdown in health—Visits Ireland and Scotland—Albion Mill—Foreign privileges—Centrifugal governor—Darwin's panegyric—Argand's lamp—Pitt's taxation—Cornish Metal Company.

THE year 1780 or thereabouts saw Watt's engine firmly established in Cornwall and elsewhere as the best means of draining mines, supplying towns and replenishing the upper stretches of canals with water, but in essentials the engine did nothing more than work a pump-rod up and down, in other words, effect reciprocating motion. Boulton wanted it to do what he had envisaged from the first, and that was to effect rotative motion at any place and to any desired extent. No one knew better than Boulton from his own experience how pressing this need already was for industry and no one realized better the potential need. He had been harping on the subject with Watt for more than a couple of years when he wrote to Watt (B. and W. Coll. 1781, June 21): "The people in London Manchester and Birmingham are *steam mill mad.* I don't mean to hurry you but I think...we should determine to take out a patent for certain methods of producing rotative motion from... the fire engine." In another letter he says: "There is no other Cornwall to be found and the most likely line for the consumption of our engines is the application of them to mills which is certainly an extensive field."

His reference in this letter to "no other Cornwall" was made because Watt had been discussing the Fen country as a possible field for exploitation, now that Cornwall was fully supplied. Boulton, however, knew that there was no money in the Fens. Watt either had not the same vision as Boulton or else he did not like facing the problem of the rotative engine. Unconscious of it possibly, he was on the threshold of a period of extraordinary intellectual output, indeed he was at the zenith of his powers. The fact that his mind was free and no longer harassed by money matters contributed greatly to this result; yet it would seem that he always needed someone behind him to push him on.

Two courses were open, either to apply steam direct to a wheel in a similar way to that which Watt had already schemed, or else to employ the common crank. The latter was the most feasible course, but the way was barred by a patent taken out in 1780 by James Pickard. The story is too long to go into here, but the partners decided neither to contest the validity of the patent, which was certainly doubtful, nor yet to take out a licence from Pickard for the use of his patent, although he offered to grant one in return for the use of Watt's condenser. Accordingly Watt set his wits to work and devised a number of substitutes for the crank. These and sundry other mechanisms he patented in 1781 (Oct. 25, No. 1306): of these substitutes the only one that was used in practice was the sun and planet gear. This is to be seen on the engine illustrated on Pl. VII, and it was characteristic of Boulton and Watt engines even later than the date of the expiration of Pickard's patent.

Watt followed this up with another remarkable omnibus patent (1782, Mar. 12, No. 1321), in which he included the double-acting engine; the double engine, that is two engines working side by side; a toothed rack on the piston rod gearing with a segmental rack on the end of the beam; and lastly a steam wheel or rotary engine. It will be realized that if the engine is to be double-acting, i.e. it has to push as well as pull, the former flexible chain connection of the piston to the beam is impossible. The problem was to transmit motion positively from a piston moving in a straight line to a beam moving in the arc of a circle. Hence the patent included the rack and sector which will effect the transmission of motion of the piston to the beam on both the up and down strokes. As may be imagined, this mechanism was clumsy and noisy in action, so much so that it was soon discarded in favour of the parallel motion which Watt included in a yet further patent of 1784 (Apr. 28, No. 1432). This likewise was a most comprehensive patent, for it described sundry other improvements, the principal of which were a steam tilt hammer and a steam road carriage.

The parallel or straight line motion was an arrangement of links involving only pin joints, and therefore well within the capabilities of the artisans of the day. It need hardly be remarked that nowadays we should effect the change to straight line motion by guides, but in those days the preparation of such plane surfaces as guides was, in the absence of the planing machine, impracticably costly and therefore not a feasible proposition. It was of the parallel motion that Watt said in his old age he was prouder than of any other invention he had made.

Although he was making all these improvements to fit the engine to work rotatively, he was still dubious whether the return they might get from making them would be worth while. Boulton kept reassuring and encouraging Watt. The latter admitted (1782, Nov. 28):

There is no doubt that fire engines will drive mills, but I entertain some doubts whether anything is to be got by them, as by any computation I have yet made of the mill for Reynolds [i.e. one that had just been ordered] I cannot make it come to more than £20 per annum which will do little more than pay trouble.

Boulton's reply (1782, Dec. 7) was:

You seem to be fearful that mills will not answer and that you cannot make Reynolds's amount to more than £20 a year. For my part I think that mills, though trifles in comparison with Cornish engines, present a field that is boundless and that will be more permanent than these transient mines, and more satisfactory than these inveterate, ungenerous and envious miners and mine lords. As to the trouble of small engines, I would curtail it by making a pattern card of them (which may be done in the course of next year) and confine ourselves to those sorts and sizes until our convenience admits of more.

This proposed procedure is nothing more nor less than standardization, and we must credit Boulton with being the first to envisage it. Yet the proposal fell on deaf ears, for two years later Watt could actually write as follows (1784, June 22):

I see that every rotative engine will cost twice the trouble of one for raising water, and will in general pay only half the money.

Therefore I beg you will not undertake any more rotatives until our hands are clear which will not be before 1785. We have already more work in hand than we have people to execute it in the interval.

Watt's remark about the rotative engines costing twice the trouble of pumping engines shows that he had not paid any attention to Boulton's brilliant suggestion to systematize production and standardize sizes—each engine was still being designed specially for the duty it was to perform.

As regards trouble, Watt should have remembered that he had now got efficient help in the drawing office. Boulton had repeatedly urged Watt to get assistance, but he was difficult to please and had turned away two assistants when Boulton introduced John Southern, twenty-four years of age and well educated. Watt was rather stiff about him and would only engage him "provided he will give bond to give up music, otherwise I am sure he will do no good, it being the source of idleness". What a philistine Watt must have been! However, Southern was engaged; he became indispensable to Watt, and afterwards to the firm with whom and its successors he remained till his death in 1815.

We have dwelt at such great length on Watt's inventions—and they were of primary order—that we are apt to overlook the fact that Boulton was an inventor too, not of course of the calibre of Watt; rather we should call him a schemer or improver, in fact an inventor of the secondary rank. We have already mentioned a few of the gadgets he introduced in connection with the hardware manufacture; we must now mention a few more for the improvement of the steam engine. As soon as he came into contact with the Watt engine, he grasped its mode of action and seems to have been stimulated by it. While he was carrying on the experiments on the Kinneil engine in 1775, Watt being in London meanwhile, Boulton suggested that "some sort of meter should be annexed to it [i.e the condenser] by which one may see the ratio of vacuum for without an outward & visible sign it is impossible to judge of ye inward and spiritual grace". Accordingly he made a mercurial siphon gauge, and seems to have been pleased with the sight of the mercury bobbing up and down in the tube. He suggested blacklead dust for piston lubrication "as it works well with iron". He was sanguine about expansive working of steam, but recognized the limitations to it.

We find him frequently discussing ideas with Watt but the latter was a somewhat discouraging person, if not actually irritating, because it was a trait in his character that everything about steam that one could suggest to him, he made out he had already tried and discarded. Without doubt this was the way in which he treated Boulton's ideas, although some of them were sound and deserved pursuing further, for example, in connection with steam boilers—a subject in which Watt was interested but one that he scarcely touched himself; he always disclaimed having done more than "somewhat improved the form and adjusted the proportions" of the boiler. In 1775 Boulton suggested "copper spheres within the water" to increase the heating surface. In 1780 he suggested iron or copper fire-tubes through the boiler. In that for Wheal Busy four such tubes 20 in. diameter were introduced; hot gases passed through two of them and returned through the other two, an arrangement that foreshadows partially the Lancashire boiler of to-day. An extension of this idea occurred to him in 1781 and, as was his wont, he communicated it to Watt in a letter (B. and W. Coll. 1781, Nov. 13), as follows:

As I know how Boilers are liable to get their joints shaked by Carriage & as the loss of Steam & of heat is more sensibly felt in very small boilers, where only $\frac{2}{10}$ of a Bush^{l} p^{r} Hour of coal will be consumed, than in large ones, & as we shall probably want many small ones, I was thinking of making a Cast Iron tube of the form of the following transverse section A. Such a boiler will be absolutely steam tight & may be surrounded with Charcoal dust & a wooden case & when wanted to be clean'd, unscrew the flanch joint a & take the part bbb altogether with the double Copper tube & then the base B is left open to Clean. No fire ever touches B for it is laped up in non conductors of heat. You will perhaps say there is too little Steam room but I don't think that is an objection in small Engines that take little & work quick. I perceive I have not well discribed ye part b which takes off but I have made a little Copper modell

& I am satisfyd that part can be made objectionless & this I think will be a good boiler for small portable Engines. It may be cast 8 or 9 feet long.

Those who are familiar with the work of Richard Trevithick will be struck by the resemblance between this and the boilers constructed by him some twenty years later for just such engines as Boulton was visualizing. In Trevithick's hands the type proved satisfactory both in construction and operation. One can imagine, however, how Watt would damn Boulton's boiler with

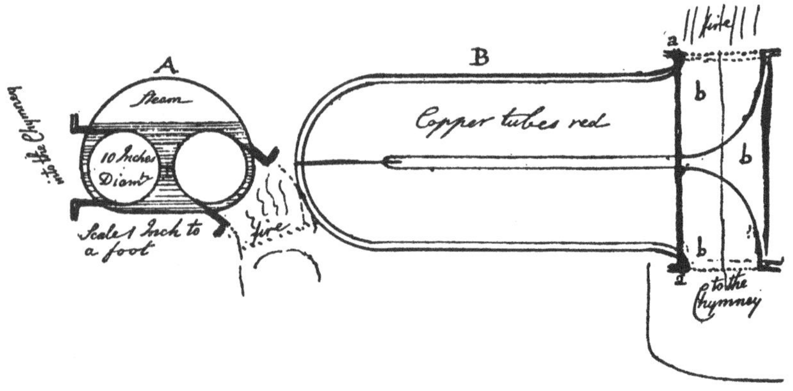

Fig. 4. Boulton's return tube steam boiler, 1781.
Boulton and Watt Collection.
Courtesy of the Public Libraries Committee, Birmingham.

faint praise and quietly turn it down. We shall learn further of Boulton's ingenuity when we come to the part of his career as a mint-master.

We must now digress for a while to tell how a great sorrow came into Boulton's life owing to the death of his second wife under very distressing circumstances on July 11th, 1783. She was found drowned in Soho Pool, the ornamental lake below the Manufactory, readily seen in the plan on Fig. 1, facing p. 42. How the accident happened was never discovered. Haymaking had been going on in the fields close to Soho House, and she had been observed walking towards the Pool. The idea of suicide was scouted, as she had been in her usual health and no reason could be assigned for such an act; the verdict of the inquest was "accidental death". Her death was a great blow to Boulton, who was passionately fond of her, as his numerous letters to her witness, letters dashed off late at night perhaps, after business was over, hardly one of them dated, but full of gentle raillery, tender messages to the family and promises of early return. One such letter written from Cornwall and, exceptionally, dated (1780, Sept. 14) we are constrained to quote, as it is typical of the rest and characteristic of the man:

So sure as there are 1728 Inches in a Cubic foot so sure was I born in that Year & so sure as there are 52 Weeks in the Year or, what you will better remember, so sure as there are 52 cards in a pack, so sure am I 52 years old this very day and yet I fear you think so little of me that you will neither have a plumb pudding for your dinner nor drink my good health....Well the older we grow the better friends I hope we shall be & though Time has a spite against Cupid's wings, yet Friendship never comes to maturity unless it hath been long nursed & cherished by old Time....

I have wrote so long a letter today to Mr Watt upon business that I have not time to indulge my self with you so long as I could wish. Let me hear from you often & tell me all you know of my dear Children who with yourself are objects of my constant prayer. I long to meet you; till then, farewell. Remember me who am ever yours & only yours

M. B.

Boulton's health was so much affected by his loss that he felt he must get away from Soho and all its business affairs. He was strongly urged to do so by Dr Darwin and by Dr Withering and, thus advised, he determined to make the effort. He decided upon a long meditated tour in Ireland and Scotland. He set out on August 22nd and went first to Dublin; he crossed to Scotland by Donaghadee to Port Patrick, travelling via Dumfries to Edinburgh. In that city he called upon Dr Black and Prof. Robison, with whom he talked a great amount of shop. From Edinburgh he went on to Carron Ironworks, a place with which he had had many associations through Roebuck, Garbett and Watt. He stayed there about a month and took the opportunity of "settling accounts", but principally of making a great number of experiments on all their iron ores. One wonders if he paid a visit to Kinneil House to look at the spot where Watt erected his first engine! He called on Lord Cochrane, ninth

Earl of Dundonald, a great inventor and schemer, who had obtained a patent in 1781 for extracting tar from pit coal, and Boulton wished to see the process which was carried out at Culross, Perthshire. In the course of his travels he made a large collection of minerals, as was his wont, and arrived home on December 11th heavy laden, greatly benefited in health, although perhaps not restored in spirits. The cost of the journey was £232. 5*s*. 6*d*., a fact that we learn from Boulton's list of travelling expenses which he always kept (B. and W. Coll.). He appears to have charged a subsistence allowance of half a guinea a day and coach fare at 9*d*. per mile. For example, the expense of his journey to and stay in Cornwall from September 14th to December 28th, 95 days at 10*s*. 6*d*. and 263 miles each way, £17. 14*s*. 6*d*., totalled £67. 12*s*. Earlier in 1783 he had been to London from February 5th to March 14th, 23 days in town at 10*s*. 6*d*. and 116 miles each way at 9*d*., totalling £20. 15*s*. 6*d*.

He plunged once again into the whirlpool of business and, after a short visit to London, went down into Cornwall to look after the engine business there. Owing to the low prices ruling of tin and copper, the greater depths to which mining had to be prosecuted and its expense, the adventurers were in a state of discontent generally. The only way of salvation that they could see was to get an abatement of the premiums on their pumping engines, forgetful of the fact that, as Watt said: "if we had not furnished the miners with more effectual means [than the common engine] of draining the water almost all the deep mines would have been abandoned before now".

The tale of unprofitable expenditure and actual losses was sufficiently alarming and Boulton, always reasonable and ready to find a *modus vivendi*, was yet annoyed by the threats held out to him by the adventurers that if he did not yield they would employ the Hornblowers to erect engines for them. The point of this threat was that Jonathan Hornblower the younger had patented in 1781 a compound engine which his faction boasted would, owing to its alleged superiority, supplant Watt's engine. The latter got really badly scared, but it turned out that Hornblower's engine embodied the separate condenser so that the partners were able to checkmate him by getting injunctions against him. Boulton wrote to Watt:

It is a disagreeable thing to live amongst one's enemies, and all the adventurers are so, except Phillips and the Foxes who are fair men, although they would rather have their engines free. I have had many hints given me that the Trumpeters [i.e. the Hornblower family] were reviving their mischief and many causes for uneasiness, but I did not wish you to partake of them, and therefore have been silent; but they are now striking at the root of us, and therefore we must defend ourselves or fall....I think if we could but keep up our spirits and be active we might vanquish all the host. But I must own I have been low-spirited ever since I have been here—have been indolent and feel as if the springs of life were let down.

This letter reads more like one of Watt than one of the usually cheery Boulton, but can we wonder at his low spirits and indolent feelings? he was still mourning over his wife's death and he had more time in Cornwall than at home to brood over it.

Watt, as we should expect, was averse from making concessions to the adventurers. He wrote to Boulton:

People do not employ us out of personal regard, but to serve themselves; and why should we not look after ourselves in like manner.... John Taylor [the Birmingham button maker] died the other day worth £200,000 without ever doing one generous action. I do not mean that we should follow *his* example. I should not consent to oppression or to take any unfair advantage of my neighbour's necessity, but I think it blameable to exercise generosity towards those who display none towards us. It is playing an unfair game when the advantage is wholly on their side. If Wheal Virgin threatened to stop unless we abated one-half, they should stop for me; but if it appeared that according to the mode settled in making the agreement, we had too high a premium, I should voluntarily reduce it to whatever was just.

Not exactly the Christian spirit on Watt's part—no turning of the other cheek—yet willing to yield a little in a legalized manner! This is exactly what Boulton did; the dues of Chacewater, for example, were reduced from £2500 to £1000 a year. Thus for the time being the difficulties were smoothed over and peace reigned for a short time in Cornwall.

The rotative engine was now a success and orders began to pour in. The first engine to be completed was supplied in 1783 to the order of Wilkinson, who always liked to be the first to secure anything new, for working a tilt hammer at his Bradley Ironworks. Not content, as Watt seems to have been, to wait for orders to come in, Boulton envisaged novel applications for the engine. One of these was to flour milling hitherto performed by wind and water mills to the entire satisfaction of the milling fraternity—they desired no change, but Boulton thought otherwise. At the Science Museum, South Kensington, among the models that came there from the firm in 1876, there are two models of corn mills arranged to be driven by rotative engines, and there cannot be any doubt that these were Boulton's schemes.

An opportunity occurred of giving an object lesson of the applicability of the engine in this novel capacity when, in 1782, proposals were set on foot to build a flour mill to serve the metropolis. Boulton was one of the prime movers in the matter, and in 1784 a charter of incorporation was sought to establish a mill known as the Albion Mill on the south side of the Thames near Blackfriars Bridge. The charter met with very determined opposition from the millers, indeed they pursued the promoters with misrepresentation and abuse. They felt their immediate interests would be affected, and the more far-sighted probably sensed the doom of the country mill that the scheme presaged and the aggregation of flour milling at tide-water that we see to-day. However the charter was granted. The erection of the building was entrusted to Samuel Wyatt, the architect. Boulton and Watt undertook to make the engines and John Rennie, the Scottish millwright, was engaged on the firm's recommendation to make the dressing and grinding machinery. It may be remarked that this was Rennie's first introduction to London, and he thus got his foot on the ladder that eventually brought him to fame as a civil engineer.

Both engines and machinery were on a scale not previously attempted and involved many new features which were the subject of much anxious thought. The first engine was started on February 15th, 1786, but it was some weeks later before it worked under load. Boulton came to town and was in attendance at the mill, where he spent days together from ten in the morning till ten at night, busy in correcting the many defects that developed. He did not allow these to engross his attention entirely but interested himself, characteristically, in the milling process itself. In one of his letters to Watt (B. and W. Coll. 1786, Apr. 15) we find this observation: "It would be a great improvement to this manufacture if we had some cheap & convent[t] means of preparing all our wheat for keeping in the bins 1st by warming and drying a very little, 2nd by cleaning it with rotative brushes & 3[rdly] by winnowing or blowing dry air through it and blowing light stuff away." If he had suggested washing the wheat as well, we could acclaim him as the precursor of all modern practice.

The Albion Mill was looked upon, and deservedly so, as one of the mechanical wonders of the day. It was beginning to prove a commercial success when it was burned to the water's edge on March 3rd, 1791, barely five years after its completion, and before the third intended engine had been installed. It was bruited about that this was the work of incendiaries instigated by the millers, but no evidence to this effect could be produced. The fact that the management had been inefficient, and that this may have led to slackness say in allowing a highly inflammable atmosphere to prevail inside the mill, would be sufficient to account for the disaster. The proprietors lost heavily, Boulton alone to the tune of £6000, and the mill was not rebuilt. It was a serious blow to him financially, but the mill had been a great advertisement for the firm and had established the economy of milling by steam power.

Contrary to the calumnies circulated by the millers, who represented the mill as a monopoly inimical to the public interest, it had an effect in reducing the price of flour in the metropolis as the following statement issued by the company and reported by several newspapers, attests:

It appears that, for the twelve years immediately preceding the completion of the Albion Mill, the average price of Wheat per quarter exceeded the average price of Flour per sack only 5s. 0¾d. and that in the five years which the Mill worked the average difference was 8s. 4¾d., which incontrovertibly proves that the operation of the Mill produced a saving of 3s. 4d. upon every sack of Flour consumed in the London Market.[1]

The sale of the rotative engine exceeded even Boulton's sanguine expectations and, seeing that there was a field for it abroad, not of course as wide as in Great Britain, he began trying to secure protection for it by privileges or patents. We have evidence of such attempts in France and the States of Holland. Out of the first-named attempt arose an official invitation to the partners to visit Paris to inspect and report on, with a view to replace it, the machine of Marly, that megatherium of ancient mill work constructed in 1676, that supplied the palace of Versailles with water from the Seine.

The partners were received and entertained most politely, It is said that Boulton dressed himself up for these occasions in the latest fashion and wore a sword, but one cannot imagine Watt doing so. Boulton certainly could carry it off, for he was a fine figure of a man and his likeness to Louis XIV was very marked. The visit did not result in any engine business, partly because of the fall of M. de Calonne, the King's minister, and partly because the shadow of Revolution was already over the land. But the visit was destined to have a far-reaching effect in another direction—that of coinage about which we shall have much to say later.

Rotative engines were erected at first in the same way as pumping engines, but later the firm undertook the entire supply and erection. The method of charging for them was different from the beginning. In the case of the pumping engine the charge or premium was, as we have said, based on the savings effected as compared with a common engine doing the same work; in the rotative engine the analogous case was the number of horses that the engine replaced; hence this number became the measure of performance of the engine. An annual premium based on the horses' power that the engine could exert was the

[1] Cf. *The Diary or Woodfall's Register*, June 23rd, 1791.

PLATE VII. THE "LAP" ENGINE, 1788, SHOWING THE "SUN AND PLANET" GEAR

Courtesy of the Science Museum

obvious way of payment. This was fixed at £6. 6*s*. in London and £5 in the provinces for each horse power.

The demands for power in the Manufactory had increased to such an extent that Boulton had an engine constructed there in 1788 for use in driving the laps or polishing buffs; hence it became known familiarly by the workmen as the "Lap" engine. Fortunately it has been preserved and is now to be seen in the Science Museum, South Kensington (see Pl. VII). It may be taken as typical of the double-acting engine of this period, with its wooden beam, connecting rod and framing, its parallel motion and its sun and planet gear.

A point of great interest in connection with this engine is that it was the first to which Watt applied the centrifugal governor, another of his beautiful inventions. The genesis of the device and how the invention came about is an interesting story characteristic of both Boulton and Watt. It is now usually admitted that a centrifugal mechanism of some kind was in use in 1788 for adjusting the distance apart of millstones, and that it was then a novel application. Now Boulton, on one of his visits to town, had called as usual at the Albion Mill to look after the partners' interests. There he noticed that a number of novel devices had been introduced. One that struck him forcibly was a device, as he explained in a letter to Watt[1],

> for regulating the pressure or the distance of the top mill stone from the Bedstone in such a manner that the faster the engine goes the lower or closer it grinds and when the engine stops the top stone rises up & I think the principal advantage of this invention is in making it easy to set the engine to work because the top stone cannot press upon the lower untill the mill is in full motion; this is produced by the centrifugal force of 2 lead weights which rise up horizontal when in motion and fall down when ye motion is decreased by which means they act on a lever that is divided as 30 to 1 but to explain it requires a drawing.

No doubt Boulton on his return sketched the device. Now Watt never needed more than a suggestion like this to set his brain to work, and he seems to have turned over the idea to

[1] B. and W. Coll. 1788, May 28.

such good effect that in December he and Southern had designed the centrifugal governor and had made one which they experimented with successfully on the Lap engine which was conveniently available. The apparatus consists of two balls attached by links to a vertical shaft driven from the engine so that, as its speed increases, the balls fly out, and in doing so lift a sleeve on the shaft. A lever with its end embracing the sleeve actuates at the other end a throttle valve in the steam pipe supplying the engine. The supply is thereby diminished, hence the speed of the engine is reduced till equilibrium is re-established.

The governor proved quite capable of the duty expected of it and in a couple of years' time became quite well known. Anyone could use it seeing that Watt did not patent it; probably he thought a patent might be invalidated if the Albion Mill's governor were cited as an anticipation as it could have been.

We have now reached the stage when Watt's rotative engine with all its improvements was fully launched on the world. Its merits began to be recognized and appreciated. In order to show this, we cannot do better than quote Darwin's panegyric on the engine and on what it was capable of doing. As we have intimated already, he was an enthusiastic admirer of the work of Watt and of Boulton, and he introduced the theme of the steam engine into his poem, *The Botanic Garden*. While composing it he was in correspondence with the partners so as to make sure of his facts. Even if Darwin's poetic merit be challenged, at any rate as a description the poem is correct except that "Newcomen" should be substituted for "Savery" in the second line. It runs as follows:[1]

> NYMPHS! YOU erewhile on simmering cauldrons play'd,
> And call'd delighted SAVERY to your aid;
> Bade round the youth explosive STEAM aspire
> In gathering clouds, and wing'd the wave with fire;
> Bade with cold streams the quick expansion stop,
> And sunk the immense of vapour to a drop.—
> Press'd by the ponderous air the Piston falls
> Resistless, sliding through it's iron walls;

[1] [Darwin, Erasmus], *The Botanic Garden*, Part I, 1791, p. 62.

Quick moves the balanced beam, of giant-birth,
Wields his large limbs, and nodding shakes the earth.
The Giant-Power from earth's remotest caves
Lifts with strong arm her dark reluctant waves;
Each cavern'd rock, and hidden den explores,
Drags her dark coals, and digs her shining ores.—
Next, in close cells of ribbed oak confin'd,
Gale after gale, He crowds the struggling wind;
The imprison'd storms through brazen nostrils roar,
Fan the white flame, and fuse the sparkling ore.
Here high in air the rising stream He pours
To clay-built cisterns, or to lead-lined towers;
Fresh through a thousand pipes the wave distils,
And thirsty cities drink the exuberant rills.—
There the vast mill-stone with inebriate whirl
On trembling floors his forceful fingers twirl.
Whose flinty teeth the golden harvests grind,
Feast without blood! and nourish human-kind.

At this point it is convenient to pause for a moment to give an account of a matter which might have had an important bearing on Boulton's future business career, but owing to outside circumstances did not lead to anything. Nevertheless, the story reveals some characteristic reactions. We refer to his connection with the Argand lamp.

One of the drawbacks to life up to the end of the eighteenth century must have been the inefficiency of artificial lighting and nowhere can it have been felt more acutely than in the home. Candles were almost the only means of lighting available, and their light went off rapidly as the wick got exposed and did not get consumed. To obviate the resulting smoke and smell, the wick had to be snuffed. Few persons realize that the impregnated plaited wick that ensures the wick curling over and getting consumed was not invented till about 1830. Oil lamps were little better, as the wick got choked, resulting in similar horrible smoke and smell. Anything therefore that improved artificial lighting was sure of a welcome. Such was the oil lamp invented by F. P. Aimé Argand (1755–1803), physician of Geneva. He made his wick tubular so that air could reach the inside as well as the outside, and he provided a glass chimney to promote the upward draught, thereby increasing combustion and the emission of light. This, it may be remarked, is the greatest single improvement that has been made in the lamp, and it has led to many other applications. Argand came over to England in 1783 and appears to have got into touch with Boulton through a Mr Parker. Argand patented his lamp in 1784 (Mar. 12, No. 1425) and on May 1st he announced his intention to commit to Boulton's care the manufacture of most of the parts of the lamp. The improvement being one that was sure to be in universal demand once it was known, and the lamp itself being easily made, it was just the thing that Soho could and should take up. The very ease of the manufacture of the lamp was its undoing, however, for the market was flooded with infringements, the tide of which Argand vainly attempted to stem. His opponents challenged the validity of the patent on the ground of prior disclosure: unfortunately it was possible to prove this, with the result that the patent was declared invalid, as the following letter of Boulton to Watt (B. and W. Coll. 1786, Feb. 24) reveals:

The Judge & jury have this day given all the Merit of the Lamp to Argand but at the same time have given a Verdict against him in consequence of Maggelan & another proving that ye invention was introduced into Engl^{d} a few days before ye date of his patent although Magg^{ln} said he could not answer precisely to dates as he had lost his first letter from Paris.... Poor Argand is oppress'd with grief.... Some of the Jury express'd their reluctance at giving an Evidence [i.e. verdict] against Argand but bow'd obedience to the letter of the Law. The Verdict was most certainly not just.

In a subsequent letter (1786, Mar. 15) Boulton continues:

Tyranny & an improper exercise of power will not do in this country. It is most people's opinion that if Parker had sold the lamps at moderate prices & had permitted Bayley, Blades & other neighbours who apply'd for them to have sold them upon the same terms as the Lamp Co. allowed Clark, there would have been no opposition to ye patent but when these men found all their glass customers running to Parkers & that they could not have any lamps to sell, they then consulted, subscribed & determined to find some means or other to overturn it either by fair or foul means. I also think Arkwright split upon this rock. If he had been a more civilized being & had understood mankind better he would now have enjoyed his patent. Hence let us learn wisdom by other men's ills.

The verdict, although in accordance with law, was decidedly unjust; it was a crushing blow to Argand and it must have been a heavy one to Boulton—he had dropped some money in promoting the patent—because the sole right of manufacturing these lamps would have been immensely lucrative to both himself and Argand. As the second letter shows, the verdict cast a red light on the uncertainties of property in patents, a light that was not lost on Boulton, for later on writing to De Luc (1787) he said:

It [the verdict] was hard, unjust and impolitic, as it hath (to my knowledge) discouraged a very ingenious French chemist from coming over and establishing in this country an invention of the highest importance to one of our greatest manufactures. Moreover, it tends to destroy the greatest of all stimulants to invention, viz. the idea of enjoying the fruits of one's own labour. Some late decisions against the validity of certain patents [he is here referring to Arkwright's] have raised the spirits of the illiberal, sordid, unjust, ungenerous and inventionless misers who prey upon the vitals of the ingenious, and make haste to seize upon what their laborious and often costly application has produced. The decisions to which I refer have encouraged a combination in Cornwall to erect engines on Boulton & Watt's principles contrary to the Law of Patents and the express provisions of an Act of Parliament; and this they are setting about in order to drive us into a court of law, flattering themselves that it is the present disposition of the judges to set their faces against all patents.

Boulton, shortly after, *did* make Argand lamps. A circular dated 1797, November, has been preserved giving "Directions for using Argand & Co's Patent Lamps, Made & sold by Matthew Boulton at Soho near Birmingham": the Manufactory indeed produced many attractive lamp designs in Sheffield plate, etc.; but naturally Boulton was only one among a crowd of other makers.

It is sad to have to relate that Argand—after setting up a lamp factory at Versoix near Geneva in 1787, which came to grief during the French Revolution—died in great poverty, where is not known—an instance only too easily paralleled of the way the world treats its great inventors.

We indicated at the outset of this volume that Boulton was always much immersed in his own affairs and the running of his beloved Manufactory, so that he had little time to devote to public work. It was not that he was unaware of the course of political events, since he mixed with every class of the community, but that he was happiest at home. Perhaps the incident that first brought him out of his shell was when the town was preparing a loyal address to King George III on the occasion of the overthrow of the Duke of Portland's administration at the general election of 1784. William Pitt became prime minister and great things were expected from him. He disappointed the manufacturing interests, however, by his proposals to levy taxes on raw materials, to the tune of one million pounds a year. The manufacturers were up in arms, convinced that the measure was suicidal. As Boulton exclaimed, in a letter to Thomas Wilson, (1784, Dec. 16): "Let taxes be laid on luxuries, upon vices, and if you like upon property; tax riches when got and the expenditure of them, but not the means of getting them: of all things don't cut open the hen that lays the golden eggs."

Boulton, who was recognized locally as a doughty opponent of these taxes, was invited by Pitt as a representative to interview him on the subject. Boulton availed himself of this opportunity to impress upon the minister the necessity of reciprocal tariffs against other nations. Writing to his friend Garbett, Boulton says:

> There is no doubt but the edicts, prohibitions and high duties laid on our manufactures by foreign powers will be severely felt, unless some new commercial treaties are entered into with such powers. I fear our young Minister [he was then only 25 years of age] is not sufficiently aware of the importance of the subject and I likewise fear he will pledge himself before Parliament meets to carry other measures in the next session that will be as odious to the country as his late attempts.

How familiar it all sounds! The anticipated happened, and Boulton joined Wedgwood in organizing opposition to Pitt's taxes. They met in February, 1785, and wrote to their friends in all quarters with the result that a Chamber of Manufacturers deliberating in London was formed. A printed statement of its objects has been preserved; generally speaking, its aims were to oppose taxes on raw materials and export duties on manufactures. Moreover, as the Irish Parliament was imposing tariffs on English goods while Irish products were let in free over here, the Chamber opposed this legislation also. Watt did not approve of his partner divagating into politics, and was in favour of leaving the agitation to others and consolidating their property—rather a selfish and shortsighted view. However, Pitt dropped the duties on raw materials, and made considerable abatements on the Customs duties. The Irish proposals too were dropped.[1]

We have mentioned above that Boulton and Watt were induced by force of circumstances to become adventurers in the Cornish mines themselves either by taking a share to secure an order for an engine or accepting a share instead of premium payments. By the end of 1780 the firm held shares in five mines, and it became their practice to take an interest in the mines. It is to be remembered that the mines were worked on the Cost Book system,[2] which did not necessitate subscription of capital. Had it been otherwise, the partners could not have become adventurers, seeing that they themselves were hard up for capital.

One of the results of this policy was that Boulton was able to effect improvements in the management of the mines. But the worst trouble was not bad management, it was rather the steady fall in the price of ore, due largely to the competition of the famous Parys Mine in Anglesea, where the ore was won cheaply in open workings. There was difficulty also with the smelting firms. Boulton thought that the only way to remedy this unsatisfactory state of affairs was to form a company to "maintain

[1] For ampler information we must refer the reader to Roll, E., *Early Experiment in Industrial Organization*, 1930, pp. 137–39.

[2] Cf. Dickinson and Titley, *Richard Trevithick, the Engineer and the Man*, 1934, pp. 7–9.

and keep the price of copper ore at a proper standard". Even as a manufacturer, a steady if high price for his raw material was less objectionable than violent fluctuations. He brought in his friends with the result that the Cornish Metal Company was constituted at Truro on September 1st, 1785, with a capital of half a million pounds. An arrangement was made with the principal mines to buy the whole of the output of ore at paying prices over a period of eleven years. Thomas Williams, the manager of the Parys Mine, was also brought in; likewise the smelters, who agreed to smelt the ore at settled prices.

The perspicacious reader will see in this company quite a good example of the trust or cartel which is usually believed to be a modern idea.

Great difficulties followed because the enhanced price of copper stimulated production. Since a profitable market could not be found, stocks accumulated; foreign competition also was encountered. The working of the scheme did not bring salvation to the mining industry, agreements had to be rescinded and all proposals for curtailing production met with determined opposition from mine owners and miners, leading up to a series of riots between 1787 and 1789. As is usual, a scapegoat had to be found and Boulton was singled out; indeed on one occasion he nearly suffered personal violence. The Metal Company closed down in 1792, at the end of the determined period, but it cannot be said to have been a success.[1] The weaker mines had to go to the wall and gradually the market recovered.

It is not unlikely that his dealings in the copper market and his efforts to find new outlets for the metal may have directed his attention to its extended use in coinage. This is the subject-matter of the next chapter.

[1] The history of this attempted cartel is detailed in Roll, E., *loc. cit.* pp. 90–4, *q.v.*

PLATE VIII. MATTHEW BOULTON, *aet.* 64

From the oil painting by C. F. von Breda in the possession of
The Institution of Civil Engineers

CHAPTER VII

COINAGE AND SOHO MINT

State of the coinage—Counterfeiting—Boulton's ideas—Appears before the Privy Council—Applies steam power to coinage—Patents coining press—Erects Soho Mint—Work as a medallist—Coinage of Great Britain—Fits out Royal and other Mints—Copper sheathing for ships.

NOW that the rotative engine was definitely off his hands and in production Boulton was only concerned to find further uses for it. We have already given in detail two such applications, but the fact was that there was scarcely any industry to which it could not and was not being applied. One application that had not been made hitherto was to coining money, which was still carried on by hand.

The subject of coinage in general had been long in Boulton's mind, because of the situation existing in this country where the coinage was in a state that can hardly be credited. It is necessary to go back a century or more to explain the reason for this, and how it was that the art of coining was in such a backward state. In this country, although not so late as in France, money was produced by "hammering", that is a die with the obverse and another with the reverse of the coin were set opposite one another in clips resembling a pair of tongs, the blank placed between them and the dies hammered on to the blank by brute force. Obviously this was a most laborious and expensive method, and so inaccurate in register that the coins were easily counterfeited. This hammered money was not superseded in England till 1662.

The application to coinage of the fly-press, already fully explained, constituted a great advance in that the two faces of the coin could be registered correctly and more accurate impressions, less easily counterfeited, could be produced. However, the press required man power no less than previously, and one of the sights of the mints of those days was to see two or three men with ropes attached to the heavy ball on the lever fixed to the screw, giving impetus to it by a rapid jerk. Our illustration of a

Mint in 1750 (Pl. IX) shows clearly how the press was operated, although it omits to show how the blanks were fed to the dies (usually by a boy). The illustration shows also the lever shears for cutting out the blanks from the metal sheets. It is not altogether to be wondered at that, in the Royal Mint, which alone had the privilege of issuing legal tender, coining should have been largely confined to the more valuable metals—gold and silver—and that the coining of inferior metals should have been neglected. To have dealt effectively with the whole of the coinage would have necessitated a very large staff. The result of this neglect was that small change was scarcely to be had, and the defect was more and more severely felt as the industrial system developed, bringing with it wage payments and a cash nexus between men. The strange thing is that the authorities at the Mint did not take proper steps to remedy the deficiency of copper coin, but apparently no attempt was made even to ease the situation and it amounted to a scandal. So great indeed was the shortage that, in despair, tradesmen and others had tokens made for them. This was done on an extended scale, indeed it is said that between 1648 and 1672 over 20,000 issues of these tokens were made.

John Evelyn in his *Diary* mentions this widespread practice when he says that tokens were issued by every tavern "passable through the neighbourhood though seldom reaching farther than the next street or two". Nearly every local museum has a collection of these tokens, showing how prevalent was this form of currency. But this was not all, for the shortage of "coppers" led to counterfeiting of what money was issued officially and, in the middle of the eighteenth century, this became an almost intolerable abuse in spite of the most stringent laws against it, the penalties for the crime being transportation—possibly hanging—often accompanied by brutal treatment. It was estimated in 1753 that half the copper coin in circulation was counterfeit. It was not this coin alone that was counterfeited, but silver and gold likewise, so much so that no business man stirred abroad without a pocket guinea balance wherewith to weigh coins whose authenticity he doubted or which he suspected had been "sweated".

PLATE IX. THE ART OF COINING IN 1750

From the *Universal Magazine*

No regal copper coinage was issued between 1754 and 1770; then there was an issue of halfpence but that was quite inadequate. In 1791 there was an issue of farthings, also inadequate. The obvious course of removing the cause of an abuse is nearly always the last thing thought of and so it was in this case; the issue of a sufficiency of genuine coin would have stopped the issue of base coin. In the matter of counterfeit coin, Birmingham, owing to the proficiency of its artists in die sinking and die stamping for buttons and the like, was easily first, so that counterfeiting became an all but recognized industry there.

Boulton's attention had been turned to the state of the currency in general, at least as early as 1772, for in a letter (Nov. 10) to Lord Dartmouth, at that time Colonial Secretary, Boulton says: "I am sure the present state of Coinage is a proper subject for the consideration of it [i.e. the Legislature] as well as the present State of Commercial Credit, there being neither coin [n]or paper sufficient in the circle of Commerce from the present state of it."

Watt states that Boulton had many conversations with him about 1774 on the subject of applying steam power to coining. We must credit Boulton with having been the first to conceive this idea and to carry it into effect. As to the coins themselves, his idea was that their intrinsic value and their accuracy of execution should be so great that counterfeiting would not be worth while. He had other ideas too, e.g. that the coin should be struck in a collar to keep its diameter exact, that the border should be raised so as to take the wear, and that the thickness should be a definite amount. By passing the coin through a gauge of the diameter and thickness of the coin and then weighing it, the purity of the coin would be determined. Supposing the coin to be made of gold and that it passed through the gauge but was then found to be underweight, it was certain that it contained base metal; or if it were of the right weight it would then be too thick to pass through the gauge. Boulton made a proof guinea embodying his ideas, but his suggestions met with no response.

Undeterred by the apathy of those in high places, Boulton proceeded to carry his ideas into practice, and he erected some coining presses of the best design then available. With this plant he was enabled in 1786 to execute successfully an order for copper coinage amounting to 100 tons for the Honourable East India Company.

When the partners were in Paris in 1786 to advise about the Machine of Marly, as already mentioned, they visited, among many other places of technical interest, the Hôtel des Monnaies or French Mint. Here they saw, among other objects, some improvements in coining that had not then been adopted by the French government. One of these was a crown piece executed by Droz[1] and struck in a collar split into six parts which were brought together at the moment of striking the blank, thus forming the edge. It will be realized that a collar has to be made in three parts at least in order that the coin shall be released after striking. Making the collar in six parts ensured easier release and enabled an inscription to be struck on the edge. If a coin or medal having an inscription on the edge is examined, faint lines showing the junctions of the part of the collar will be observed.

Besides this split collar, Droz "had also made several improvements in the coining press and pretended to others in the art of the multiplying of the dies", so that Boulton was greatly impressed. From the beginning Boulton's aim had been to improve the national coinage. He brought to bear on the Government the influence of his friends, assisted by petitions from merchants, traders and others to bring about a reform of the crying evil of counterfeit coin. At length in 1787 a Committee of the Privy Council was appointed to take the matter into consideration. The author was at some pains to find the minutes of this Committee and was rewarded by finding at the Record Office a part, but only a part, of them: "Papers relating to copper 1798–1802",[2] concerned with the work of the

[1] Jean Pierre Droz (1746–1823), a celebrated French die sinker, *see* Forrer, L., *Biographical Dictionary of Medallists*, 6 vols. 1904 onwards, *art.* Droz.

[2] Record Office, Board of Trade 6/117, 118.

"Right Honourable the Lords of the Committee of Council appointed to take into consideration the state of the Coins of the Realm", to give it its full title. Much use of these papers has been made below.

In a memorial to this Committee, dated 1790, May 28th, Boulton gives a résumé of what happened. In December, 1787, he was invited to appear before the Committee and on January 15th following he did so and was examined as to his proposals for preventing counterfeiting. He expounded his views with such conviction that he was authorized to prepare and submit to the Committee patterns of halfpenny pieces on his new plan. These "for the sake of expedition he was obliged to have struck by the Press Mr. Droz had constructed at Paris, there being no one in England capable of striking such". The obvious reason of this was that they had an ornament on the edge and this could only be struck in a split collar. In May and June following he presented these pattern pieces, and he also made and presented some with the inscription on the edge: "Render unto Caesar the things that are Caesar's." The latter were struck in "a new Press constructed by your Memorialist at Soho". He answered questions as to the rate at which coins could be struck. He was "induced at great expence to exert himself in advancing the execution of his own Inventions and all the Machinery necessary" which took up much time. "He was influenced", he continues, "by the ambition of making more excellent coin than had ever been seen, and establishing an effectual check upon those who counterfeit it." He goes on to say that he "engaged Mr. Droz at a very great expence to engrave the original puncheons and matrices for the proposed copper coinage and to superintend the execution of it."

Boulton further offered to execute the halfpenny coinage at a cost not exceeding half that incurred by the Mint in producing the coin then current. Although the patterns were approved, a delay ensued that is almost inexplicable unless we attribute it, as was believed to be the case, to the intransigeance of the officers of the Mint who would not admit their incompetence, with the old-fashioned and inadequate machinery at their command, to execute nearly sufficient coinage for the rapidly growing needs of the community. This is borne out by this restrained statement of Watt: "His proposals on this head were not however approved of by those who had the management of His Majesty's mint & there the affair rested for the time."

Meanwhile, however, Boulton was so elated at the prospect of having the new coinage to execute that he felt he must make preparations for it. He had felt dissatisfaction with the presses already set up because they were violent and noisy in action, and he determined not only on new presses, with all the latest improvements, but on applying a steam engine to drive them and erecting a new building, afterwards known as Soho Mint, to accommodate the plant.

By the end of 1788 he had fitted and set to work six presses. Coupled, of course, with this was the provision of plant for rolling the copper ingots into strip or sheet to close limits in thickness, annealing and scaling the strip, cutting out the blanks, rumbling them to remove the arris of the edges, and bagging them, not to speak of forging and multiplying the dies. In all this, it may be admitted, he was only applying known methods, but with greater accuracy and with numerous improvements in details. He did more than this, however, for he organized the work by spacing each machine relative to the rest and feeding the material from one to another machine in boxes through shoots so as to cut down handling to a minimum. Thus equipped, he executed a copper coinage for the American colonies and a silver one for the Sierra Leone Company.

Concurrently with this practical work, Boulton continued to urge his views upon such members of the Government as he had access to. For example, writing in 1789, April 14th, to Lord Hawkesbury, later 1st Earl of Liverpool, who had been Master of the Mint in 1775 and President of the Board of Trade in 1786, Boulton said:

> In the course of my journeys I observe that I receive upon an average two thirds counterfeit halfpence for change at tollgates &c, and I believe the evil is daily increasing as the spurious money is carried into circulation by the lowest class of manufacturers who pay

with it the principal part of the wages of the poor people they employ. They purchase from the subterranean coiners 36 shillingsworth of copper (in nominal value) for 20 shillings so that the profit derived from the cheating is very large.

This shows that the situation with regard to bad money must have been steadily getting worse since 1753, when, as we have said, it was "fifty fifty". Boulton's statement as to the injustice done to workmen who were too poor to resist the imposition upon them, is corroborated by a paragraph in the local press:[1]

> The advantage taken by certain individuals of poor workmen in this place loudly calls for the interference of the officers to put a stop to the circulation of counterfeit copper coin. On Saturday night last a poor workman who, upon an average does not get more than 9*s*. or 10*s*. per Week had halfpence to the amount of 5*s*. forced upon him, in part of his week's wages, which were not intrinsically worth more than 2*s*.

In the letter to the Earl of Liverpool quoted above, Boulton goes on to say:

> The trade is carried on to so great an extent that at a public meeting at Stockport in Cheshire in January last, the magistrates and inhabitants came to a resolution to take no other halfpence in future than those of the Anglesea Company and this resolution they have published in their newspapers.

Still no alleviation of the situation was afforded as should have been expected from the Government, with the result that local bodies, firms and private individuals were driven to issue coinages of their own. These were not in any sense counterfeit, being of full weight and fineness, as we shall see; in some instances they were current over a wide area.

Upon the production of these coinages Boulton entered with his accustomed energy, and he had now an added reason for doing so—they served as an outlet for the copper being produced in the mines of Cornwall and Anglesea, an account of which we have given in a previous chapter. It would take up too much space to enumerate the localities for which and

[1] Aris's *Birmingham Gazette*, January 30th, 1786.

individuals to whom these tokens were supplied; perhaps the most interesting in design were the ½*d.* tokens supplied to his business friend, John Wilkinson, 1787–93. Of these there were a large number of issues. On the obverse they had the bust of Wilkinson, on the reverse was a tilt hammer; a female figure with a cog wheel and a boring tool; a smith striking on an anvil with a brig in the background; a crown and Welsh harp; or a figure of Justice. The emblems on the reverse were repeated on tokens struck for other persons. The quantities required were very considerable. On December 8th, 1790, Wilkinson writes: "I shall be content if I can have about 5 tons more [i.e. of the halfpence] speedily: a further quantity will be wanted for 1791." Besides these copper tokens, Wilkinson had a silver one, value one-sixth of a guinea (see Pl. X); furthermore he issued paper money. This coinage of tokens continued till the Government bronze coinage was introduced in 1843.

Boulton was eventually so satisfied with his improvements in coining machinery that he decided to take out Letters Patent for them, which were granted to him, describing himself as an "Engineer", on July 8th, 1790 (No. 1757), under the title "Application of motive power to Stamping and Coining". Boulton says in the Specification:

> The essential parts of this Invention are—First, the applying the powers of mills or steam engines to the working the presses in place of men's labour, as has hitherto been practised. Secondly, the applying the elastic force of the atmosphere acting upon the piston of an air pump to give the necessary velocity to...strike the blow which coins the money. Thirdly that the arm H which raises the pistons of the air pumps...is totally disengaged from the...fly of the press...during the time the piece F. f acts upon it by which means that bar is allowed to recoil in a natural manner and does not give the shakes which would be occasioned to the whole machine by its vis inertiae if it were acted upon immediately....

This description will be best understood by reference to the drawings attached to the specification (reproduced in Fig. 5). The patentee states that these drawings "are made from the real machine in their just proportions according to the scales marked upon them respectively".

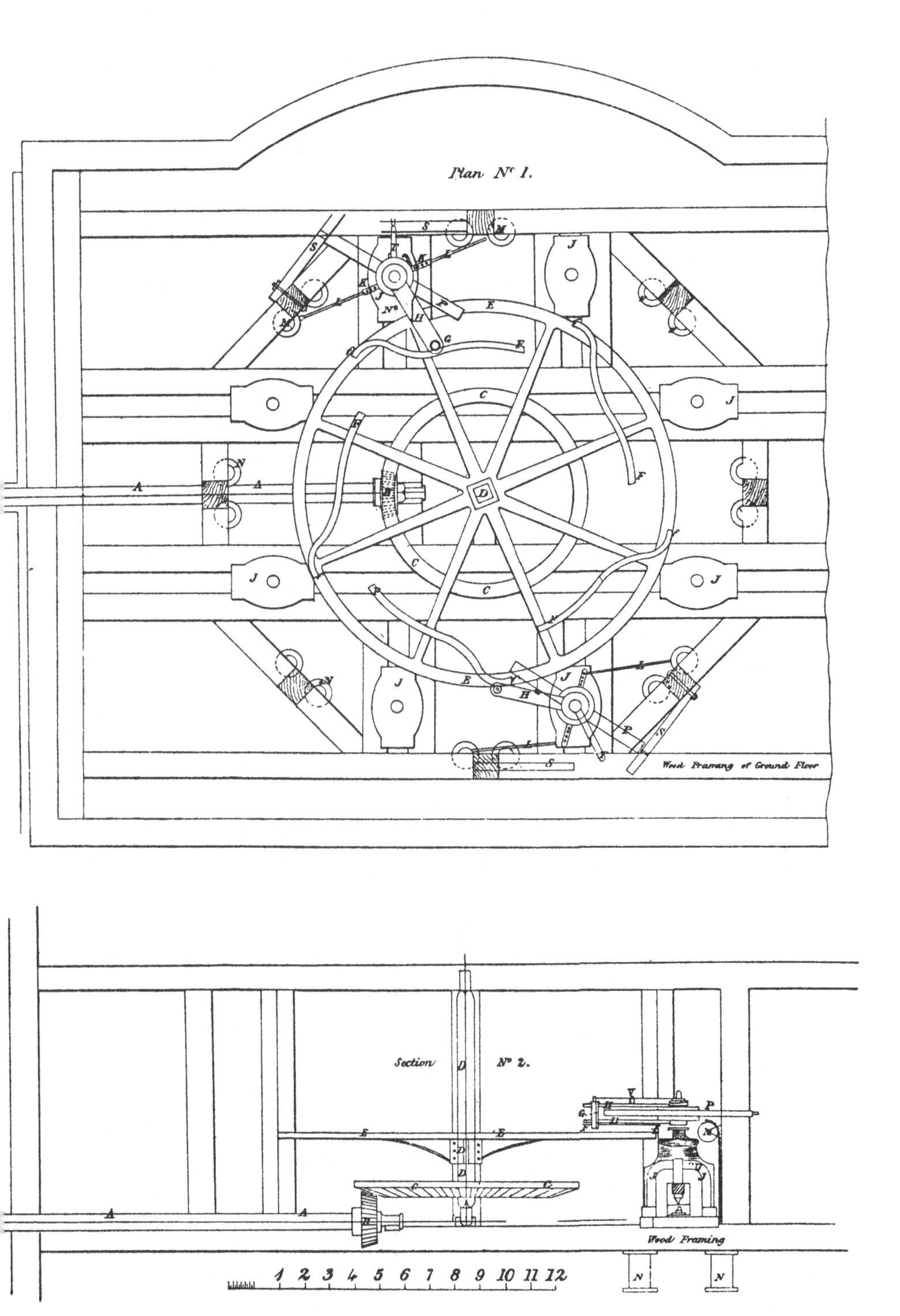

Fig. 5. Boulton's coining press, 1790.
From the Patent Specification Drawing.

It will be observed that there are eight screw coining presses *JJ* (but there might be a greater number); seen in plan and elevation, arranged in a wooden framing around a large horizontal wheel *E*, driven at 12 revs. per min. by bevel gear *BC* from the engine shaft *A*. On the upper surface of the wheel *E* are five coin plates *Ff* which engage in turn with a pin *G* on the arm *H* on the axis of the coining press *JJ* and is thereby rotated through an angle of about 60°. To this arm *H* are secured two levers *KK* which adjustably by chains running over pulleys *M* are attached to pistons in airtight cylinders *NN* below the framing. Thus the partial rotation of the arm *H* causes a vacuum in the cylinders *NN*. As the arm *H* approaches the end of its travel a detent *O* engages with a stop *R* (not readily seen owing to the small scale), on the arm *P*, which is the fly of the screw press, and locks it to the bar *H*. When the rotation of wheel *E* eventually releases the arm *H*, the air pistons being no longer resisted, rotate the fly *P* and its screw, bringing the die down on the blank below in the usual manner. The elasticity of the blow allows a spring to throw off the detent and the fly *P* rises independently of the arm *H* and all is ready for the next cycle. With the velocity stated the number of pieces coined would be 12 × 8 or 96 per min. but any one press could be temporarily put out of action. With slight alteration to the dies the press could be made to cut out blanks.

We are in a position fortunately to give a description of the Soho Mint from a MS. account dated 1792 in Boulton's own handwriting and we cannot do better than make the following extended extract from it:

This Mint consists of eight large coining-machines which are sufficiently strong to coin the largest money in current use, or even Medals; and each machine is capable of being adjusted in a few minutes so as to strike any number of pieces of money from fifty to one hundred and twenty per minute, in proportion to their diameter and degree of relief; and each piece being struck in a steel collar, the whole number are perfectly round and of equal diameter. Each machine requires the attendance of one boy of only twelve years of age, and he has no labour to perform. He can stop his press one instant, and set it going again the next. The whole of the eight presses are capable of coining, at the same time, eight different sizes of money, such as English crowns, 6-livre pieces, 24-sous pieces, 12-sous, or the very smallest money that is used in France. The number of blows at each press is proportioned to the size of the pieces, say from fifty to one hundred and twenty blows per minute, and if greater speed is wanted, he has smaller machines that will strike two hundred per minute.

As the blows given by Mr B's machinery are much more uniform than what are given by the strength of men's arms when applied to the working of the common press, the dies are not so liable to break nor the spirit of the engraving to be so soon injured; yet nevertheless from the natural imperfections of steel and other unavoidable causes, some time will be lost in changing the dies and other interruptions. However, it is decided by experience that Mr Boulton's new machinery works with less friction, less wear, less noise, is less liable to be out of order, and can strike very much more than any apparatus ever before invented; for it is capable of striking at the rate of 26,000 écus or English crowns, or 50,000 of half their diameter, in one hour, and of working night and day without fatigue to the boys, provided two sets of them work alternately for ten hours each.

From this description it is clear that each press could be stopped and started independently of the rest, no doubt by the aid of a clutch. Then again the speed of the press was adjustable, without altering the speed of the engine, from 50 to 120 blows per minute, and this was done doubtless with change gearing in a well-known manner.

How largely automatic Boulton had rendered the operation of coining is shown by the statement that the labour required was only that of boys, and even then it did not entail hard work; in other words the feeding of the blanks and the removal of the coins, when struck, were automatic. Welfare workers in factories of to-day will be intrigued to learn that these boys were dressed in white trousers and jackets, which were washed every week. It seems that, when orders were plentiful, the boys worked on the night shift and that the day was of ten hours' length. There is much of import to the social historian in the conduct of the Manufactory generally, were space available to dwell upon it. Another point of interest is that Boulton evidently had smaller coining presses that would strike 200 blows per minute; probably these were the small ones he had originally constructed.

A medal is in existence which would appear to have been struck for abroad in order to indicate the capacity and speed of the presses (see Pl. X). Its date is not earlier than 1798. The obverse has a bust of Boulton and a raised rim like the guinea already described. On the reverse is a series of concentric circles, each representing the size of a coin, and in each circle is a figure which records the number of pieces of that diameter that could be struck per minute. The legend is in French, which makes one suspect it is Droz's work. The legend is much abbreviated but may be translated thus:

Mr Boulton erected at Soho, England, in 1788 a steam engine to coin money. 1798. He erected a much superior (engine) with 8 new presses. These circles and figures mark the diameters and number of pieces struck per minute by 8 children without the least fatigue of smallest or greatest volume. Or of 8 different sizes together. The effect can be increased to the necessary amount.

The presses were capable when working to full capacity of turning out 1200 tons of coin per annum; hence the supply of such a large quantity of copper began to present difficulties. Instead of being faced with a glut, as we have indicated that Boulton was in 1785, the market, knowing his requirements, raised the price of copper about £6 per ton. Upon this the white-metal button makers lowered the wages of their workmen, alleging as the cause this rise in the price. To divert attention from themselves, a scapegoat had, as usual, to be found, and they gave out that they had to thank Mr Boulton for it. What happened and how he met the situation is best told in his own words in a letter to Thomas Wilson (1792, Feb. 26):

...From the misrepresentations that have been made by the delegates, this town has been greatly misguided and I expect every hour riots of a serious nature.

Workmen are parading the streets with cockades in their hats. They are assembled by beat of drum and headed by Ignorance and Envy, with their eyes turned towards Soho.

Yet I am no competitor with the Birmingham trades. I followed no business but what I have been myself the father of and I have done much more for the Birmingham manufactures than any other individual. I have declined the trade of White Metal Buttons which is the article so much affected by the rise of metals, and that in which the rioters are employed.

I mix with no Clubs, attend no public meetings, am of no party, nor am I a zealot in religion; I do not hold any conversation with any Birmingham person, and therefore I know no grounds but what may be suggested by wicked and envious hearts for supposing me to be the cause of the late rise of copper.

However, I am well guarded by justice, by law, by men, and by arms.

Boulton, like a good general, prepared for an attack on the Manufactory. The nature of the preparation is shown by the steps taken which consisted, so he told Watt, in enrolling his trustiest workmen, for whom he provided 200 oak cudgels, and cleaning and charging his cannon. Fortunately the danger blew over and the attack was not delivered.

Proud of his achievements in coining, Boulton took up almost simultaneously the more exacting work of producing medals. With the advice and help of Benjamin West, P.R.A., he executed a medal to commemorate the recovery of the King in 1789 from his second attack of insanity. Boulton presented the first specimen to the Queen through the intermediacy of his friend Jean André de Luc, who occupied the post of Reader to Her Majesty at Windsor. She expressed herself as much pleased with the attention.

The success attending this venture induced Boulton to prosecute further his exertions in this medallic field, not so much for the sake of profit, for many of them were issued only for presentation to individuals, but because of the credit thereby redounding to Soho and incidentally to the artistic standing of his mother country. Boulton's princely munificence was exemplified by his action on the occasion of Lord Nelson's last victory by executing a Trafalgar medal (see Pl. X), an example of which Boulton presented, with the sanction of the Admiralty, to every man who took part in the action and only to them. For officers it was in copper and for the men in pewter.

In this medallic work particularly, Boulton owed a great debt, not only to artists and sculptors such as West, Nollekins, Flaxman and Bacon, but, in particular, to two artists in die sinking, Droz and Küchler.

The first of these we have mentioned already; he did a considerable amount of work for Boulton. A series of medals commemorating scenes in the French Revolution, of note for their fine execution, is most probably by him. But as regards help in the preparation of dies and in the construction of the press itself, Droz was a disappointment, as we have already stated. Correspondence that passed between Boulton and Watt in 1790 (B. and W. Coll.) goes to show that Boulton's principal grievance against Droz was that the latter had not imparted, as had been expected, any new method of multiplying dies. It has to be remembered that a press comprising a number of coining heads will require a die for each one, and all these dies must be absolutely alike. This is effected by having a steel punch on which the obverse or reverse of the coin is engraved by the artist. This is hardened and preserved as a master punch. It is driven into a softened piece of steel which, after hardening, becomes the die used in the coining press. If the coinage is to be a long one the master punch is used only for the preparation of a few secondary master punches, so that the pristine sharpness of the original master punch may not be marred. Now this way of multiplying dies had long been known and practised in Birmingham by the button makers, so that Droz was not really helpful. We quote once more from Watt's *Memoir*:

> Mr. Droz was found to be of a troublesome disposition. Several of his contrivences were found not to answer & were obliged to be better contrived & totaly changed by Mr. B. & his assistants. The split collar was found to be difficult of execution, & subject to wear very soon when in use, & in short very unfit for an extensive coinage. Other methods were therefore adopted & it was laid to rest. Mr. Droz was dismissed after being liberally paid.

The correspondence alluded to shows that Boulton had promised Droz £1000 for instruction in the art of multiplying dies and had promised Droz, further, a sum of £700 for making accurate screws for the coining heads. As we have stated already when speaking of the fly press, this making of screws, before the day of the screw-cutting lathe, was a difficult and tedious task, requiring a high degree of skill. Whether Droz disclosed any method better than what was already in use in Birmingham we have not found out. The difference between promise and performance led to a dispute which was submitted to arbitration. Apparently this was in Droz's favour, for he appears to have got away with the sums mentioned.

Conrad Heinrich Küchler (1740?–1810) was a Flemish medallist, who was engaged by Boulton in 1793, possibly to take the place of Droz. He executed, among many other pieces of great merit, a memorial medal of Louis XVI, a companion one of Marie Antoinette, and the Trafalgar medal already mentioned. How far Boulton was indebted to Droz and to his other helpers in the contriving of the various details of his coining press, we cannot say definitely, but we can be sure that the debt was a real one. The split collar was certainly used for, as we have explained, a coin or medal with an inscription on the edge such as the Trafalgar medal, can only be struck with the aid of this device. Küchler left Soho but was employed by the firm in London up to the time of his death.

As time went on Boulton, following the policy he had adopted with "toys", recruited his staff by training them. One of these was John Phillp (1780–1820). He was apprenticed as a die sinker and was also favourably known as an artist. He cut the dies of the funeral medal to be referred to later. We mention him particularly because he was reputed, and we believe rightly, to be Boulton's natural son.

In regard to the machinery, Boulton came to rely upon his own staff, in particular James Lawson, John Bush and Bill Harrison. William Murdock, too, after his return from Cornwall, was helpful and was able to add his testimony after Boulton's death as to what the latter had accomplished, in these words:

To his indefatigable energy and perseverance in pursuit of this, the favourite and nearly the sole object of the last 20 years of the active part of Mr. Boulton's life is in a great measure to be attributed the perfection to which the art of coining has ultimately obtained.

We have now followed fairly closely the story of the realization of Boulton's aim in applying the steam engine to coining and in improving the coining press; we have now to relate how his further aim of the improvement of the national coinage was achieved. Nothing was done till 1797, in spite of petitions to Parliament and memorials to public departments, which alone could move in the matter. At length Boulton was rewarded for his persistence by an order to execute coinage to the amount of 500 tons at £108 per ton, to consist of 20 tons of 2*d*. pieces and 480 tons of 1*d*. pieces. The contract was signed on June 9th, 1797, at the Treasury, Boulton being in attendance for the purpose. The proclamation authorizing the new coinage appeared in the *London Gazette*, of July 26th, 1797. The proclamation directs what the weight of the pieces is to be (e.g. the penny piece[1] is to weigh one ounce) but says nothing about their size. There seems to be no doubt that it was to Boulton we owe the idea that our coins should serve as fairly exact measures of length as well as of weight, measures that should be available wherever a shilling's worth of copper could be collected. This is shown by Boulton's letters[2] (1797, May 19) to John Southern, written from London while the contract for the coinage was being negotiated. In this letter Boulton says:

> As I intend there shall be a coincidence between our Money, Weight and Measures by Makeing 8 two-peny pieces 1 lb or 7000 gr and to measure 1 foot; 16 peny pieces 1 lb and 17 to measure 2 feet; 32 half-pence 1 lb, and 10 to measure 1 foot; 64 farthings 1 lb, and 12 to measure 1 foot, I therefore beg that Bush would turn an original Die & an original Coller of the peny size & Strike 17 blank pieces in them to see that 17 are an exact foot & if too long or too Short let them be adjusted untill the size is exact & then let those original dies & Collers be kept as an exact Standard for the whole Coinage of pence. When that is adjusted then let the same be done for two-peny pieces (viz 1½ inches) for unless they are turn[d] perfectly exact to the same Standard my Gages will be of no use....

[1] It is interesting to note that it was only then that the copper penny became a current coin of the realm.

[2] Reproduced in the *Numismatic Chronicle*, Ser. IV, XIII (1913), p. 379.

I will also thank you to tell Mr. Foreman that I am obliged by his letter & that I have dined with Col. Fullerton. I hope you have not lost sight of the new Mint, for I am perswaded it will be far more compleat & harmonious than the present one, & the more I think of it the more I am pleased....

I have been with the Duke of Portland two hours today assisting & divising new Laws to prevent the Counterfeiting the Coin of this & all other Countries & have seen such a Collection of base money & base Arts as would astonish you & convinces me that all which Mr. Colquhoun says in his Book is true.

I am to present my Specimens to the Kg. on Monday next at Windsor but shall go on Sunday. I fear Bush is behindhand with the air pumps, the Cranks & Counter or recoil pumps at the Mint.

Writing again to Southern on May 28th, 1797, Boulton says further:

I have obtained a Standard foot from one in ye possession of the Roy^l^ Society & also a french foot which I have sent herewith p^r^ Coach.

Sir George Shuckborow is preparing to go before Parl^t^ with the subject of weights & measures but he will not be ready before next Session. I have seen his apparatus which is accurate & expensive. I will tell you ye particulars when we meet being now in haste.

As actually coined, the weights were strictly adhered to, but the diameters of the pieces varied as shown in the following table (dimensions in inches):

Piece	Dates of issue			
	1797	1799	1806	1807
Twopenny	1·6	—	—	—
Penny	1·3	—	1·35	1·35
Halfpenny	—	1·2	1·15	1·15
Farthing	—	0·925	—	0·85

It is an interesting fact not generally known that the standards thus set have been maintained since, and that our bronze coinage to-day embodies definite units of length. Our penny is 1·2 in. diam.; our halfpenny 1 in. diam.; and our farthing 0·8 in. diam.; so that 10, 12 and 15 respectively, placed edge to edge, measure

1 foot. If any one is short of a measure of length and has a supply of coppers, he can always improvise a foot-rule. This can easily be checked. The author has done so with twelve reasonably perfect halfpence and found their combined length correct to 0·01 in. +, which is near enough for most practical purposes; with newly minted coins, the accuracy would doubtless be greater.

The Duke of Portland who is mentioned in the above letter was the 3rd Duke, then Home Secretary. "Mr. Colquhoun" was Patrick Colquhoun (1745–1820), a remarkable character, at that time occupying the position of metropolitan police magistrate. He was the author of books and pamphlets correcting abuses of the day.

The thoroughness of Boulton when he took up any question is shown in the way that he secured a standard foot. He went for it to the fountain head because, as a Fellow of the Royal Society, he knew of the work that had been done there in producing standards of length by Sir George H. W. Shuckburgh-Evelyn (1751–1804) mentioned in the letter.

We may interpose here the remark that the twopenny piece, familiarly known as a "cartwheel", was so clumsy that it was quickly discarded and no more were coined after 1798. They have proved useful, however, in another direction; as a boy, the author saw this piece used as a 2 oz. weight in the marketplace of the country town where he was born.

From this point onwards we are enabled from the Coinage Papers[1], to which reference has already been made, to give in greater detail the course of events with regard to the national coinage.

In the first place we find particulars of the contract for 500 tons of coin already cited. Another contract of the same amount, similarly allocated as to denomination of coins, was entered into, for on December 24th, 1798, we have a certificate from a certain Joseph Sage, an official of the Mint, to the effect that the coining of 960 tons of penny pieces and 20 tons of twopenny pieces has been "Executed according to his [i.e. Boulton's] contract & that only 20 tons of 2d pieces remain to

[1] Record Office, Board of Trade, 6/117,118.

be coined". A further order must have been given, however, for in a letter of August 21st, 1799, Boulton states that the "total quantity of peny pieces coin[d] since June 1797" to have been "1266 Tons 18c 3qr 25lb 10oz equal to 45,407,440 of pieces".

In another letter dated December 25th, 1798, he mentions, incidentally, that "at 16 to the lb. [i.e. the penny pieces] he has only erred $\frac{1}{2425}$ too little". These statements sound too meticulously accurate but, knowing Boulton's character, we are inclined to accept them.

It will be observed that so far only twopenny and penny pieces have been coined—the halfpenny piece, the most generally used coin, had not been touched. Boulton kept on worrying "My Lords" to give an order for these coins, and he had prepared specimens which he wished to submit. Here are a few extracts from the correspondence (July 10th, 1798):

I am fully persuaded that the present evils & Inconvenience attending the Coppr Coin can only be effectually remedied by coinage of a large Coinage of half pence, in conformity to the...application I receive verbally and in writing.

Again (June 8th, 1799):

I have executed the whole of your Lordship's orders for pence and the Publick are daily writing for more; having now many orders for pence on my books, but in general they are much more pressing for Halfpence as they say they are inundated, and greatly inconvenienced with counterfeit Halfpence; which can only be suppressed by the Introduction of good legal Halfpence and I fear they will grow very discontented unless supplied soon....

By my Zeal to accomplish the Desires of your Lordships, I have incurred the Displeasure of all the mining interest of Cornwall, and thereby lost a branch of my business in that county, more profitable than coining; and whatever disappointments may befall me, they, and their Leader will exult in; which is very mortifying after taking so much pains in investigating the copper trade as well as putting myself to so much trouble and expence in improving the art of coining, so as to have little more to be done or wished for.

We have, too, an interesting "Statement of the Weight and

Value of the several Denominations of the Copper Coin issued in the years 1797 and 1798 and 1799". It is as follows:

Twopenny Piece	40 tons	8	coined out of the lb. wt. of copper
Penny do.	1226	16	,, ,, ,, ,,
Halfpenny do.	526	36	,, ,, ,, ,,
Farthing do.	26	72	,, ,, ,, ,,
	1818		

This confirms and amplifies the previous statement.

Apparently the order followed shortly, for halfpence were coined by Boulton in 1799 and these were followed quickly by farthings.

Among these official papers we come across unexpectedly an account by Boulton of private detective work and of raids on coiners' dens in Birmingham. The account is so thrilling and vividly written that we are fain to give it in full. Comment would almost seem to spoil it, but the idea of the septuagenarian Boulton in person leading his "own trusty men" fills one with admiration. Here is the story:

To the Right Honorable the Lords of the Committee of Council appointed to take into Consideration the state of the Coins of this Realm.

My Lords,

Having observ^{d} many Counterfeit pence getting into circulation in sundry parts of ye Kingdome, I have taken some pains to find a clew to the makers, and by a proper application of 50£ sterling I have obtain^{d} my purpose. In consequence thereof I took 14 of my own trusty men (for I durst not trust our Birmingham Catchpolls) with the Birm^{g} constables & their assistants (having previously obtaind the legal powers from our magistrates) & attacked 3 different Manufactories at the same instant & though I had ye precaution to conceil from the Constables &c the names of the parties untill the moment of attack yet one of them whose name is Pitt & is an old offender, found means to elude our searches, as his Buildings are constructed to that purpose; but we succeeded in our attack on the other two.

In the house and Shops of Richd Barber we found a coining press with Dies fixt on it for striking counterfeit peny pieces, of which

I send your Lordships a specimen, there were a number of Blanks as well as money ready coined out of the sd Dies, we also found a Milling Machine, Stamps & other tools adapted for the purpose all which we seizd with some counterfeit Shillings and likewise Barber himself, with some of his Children that assisted him. At the same time another squadron of my people went to the House of Thos Nichols & were informed that he was in the upper most Shop. They mounted & enterd, but it was empty. Upon observing a secret door, they attempted to pass, but found some resistance on ye other side, & a struggle ensued at length ye Constable thrust his staff through, & upon the sight of it Nicholls, like Harlequin, jumpd through a trap door upon a ladder, which he instantly kickd down and then descended into a lower room in which there is no door, & he escaped through a Window contrivd for that purpose.

There were found in his possession great quantities of countf^{t} half crowns & Shillings both Silver^{d} & unsilverd, many half pence, counterfeiting those of the present & last Reign. Many of the Shillings were wrapt up in papers of 21 each—also one Gilt Counterfeit 7s piece with Dies, & some letters & accounts which serve to show the names of the persons who are the Circulators of this base coin.

I have reason to believe there are many others in the Town of Birm^{gm}, who follow these Trades to a very great extent but they divide the operations—one man Rolls another cuts out the blanks & mills them, another coins them another Silvers them & others sink or engrave the Dies. All these people live in different parts of the Town but they form themselves into Clubs, and resort to certain Public Houses wch are known—

I think it possible to exterminate this Class of Coiners; but it will require attention, activity, silence, prudence & some knowledge of the parties: & it will require a considerable expence, for unless Men are paid well, it will be impossible to get ye necessary information, & the Catchpolls complain that they have not been paid sufficient for their time & expences. Your Lordships know enough of human nature to know that any arguments about public good or Honorable principles, stand for nothing with these men.

I have given one of them 50£ to lead me unto the Clew of detecting these people & it will require a considerable expence to compleat this business: but if I enter into that subject I shall loose the post.

I take the liberty of inclosing for your Lordships perusal a Copy of the depositions taken by Mr. Villers & Mr. Hicks our 2 Magistrates—

I also inclose an advertisement I thought necessary to publish for the apprehending of Nichols & offering 20 guineas reward. I likewise send herewith 6 pieces viz:

one Copper Silverd counterfeit half crown
one Do Do Do shilling
one Do unsilvered — Do
one blank for peny pieces } their weights is = 21 pieces in
one coind counterfeit copper peny } a lb. & were bought in Oct[r] when copper was cheaper
one counterfeit halfpeny weighing after ye rate of 100 pieces in one pound.

I have acted in this business to the best of my Judgment and I hope will not be disaproved by Your Lordships whose advice I now stand in need of & beg to be honourd with soon—

I am with the highest respect
My Lords
Your Lordships most faithfull
& most devoted humb[l] Serv[t]

Matt[w] Boulton

Soho nr Birm[gm]

Jan[y] 31 1799

The examination of the prisoners took place before the magistrates on January 30th. Barber was remanded in order to prosecute him at the assizes at Warwick on the ensuing March 25th and the case was prepared by the Treasury Solicitor. However, it was decided it would be better policy not to bring the case into court, and Barber was discharged by proclamation. He was lucky, for there is not much doubt but that he would have been found guilty and transported for life.

One further quotation must be given from these Coinage Papers. Boulton had put in a Memorial to "My Lords" praying for a continuance of the coinage of penny, halfpenny and farthing pieces, and had been told that there were many objections to his proposals. In his answer to this (Soho, 1802, Mar. 29) Boulton said:

the principal object I have aimed at both in my coinage & Memorial has been to accomplish the wishes that Mr. Pitt express'd to me some years ago viz. to render the coin so difficult to make as greatly to check or, if possible, put an end to the counterfeiting of it, which I have done...by making the intrinsic value bear a greater proportion to the nominal value; i.e. by putting 36 pieces in a Pound in the new coin instead of 46, as in the old coin, & have thereby taken away the temptation to counterfeitors.

It may be noted that a further contract between the Government and Boulton for the copper coinage of England and Ireland, pending the erection of the new Mint mentioned below, was entered into on March 26th, 1805. A paragraph in the *Globe* newspaper of June 30th, 1806, is of interest as showing that the coins were distributed not from the Mint but by the contractor who was paid directly by the customer: "The new penny pieces and farthings are issued by the contractors Messrs. Bolton, Watt & Co. (London-street, Fenchurch street) in casks weighing net 3 cwt. value £33 12*s* of which the twelfth-part are in farthings. The halfpence will not be ready for delivery these two months."

The amount of the coinage work carried out at the Soho Mint may be judged from the fact that 3531 tons of copper coin were minted between 1797 and 1808.

A somewhat analogous job was undertaken by the Soho Mint in 1809, and that was to recoin or restamp Mexican silver dollars into 5*s*. pieces for the Governor and Company of the Bank of England. These pieces were greatly admired; deliveries in London commenced during the last month of the year as shown by a paragraph in the *Globe* newspaper of January 5th, 1810:

On the 14th ult thirty-eight casks of dollars, part of those transmitted some time since from London to be stamped at the Soho Manufactory near Birmingham, amounting to £50,000 were delivered to the custody of persons appointed to receive them, and sent by Snell's Canal for conveyance to the metropolis where they arrived in the ensuing week. On the 22nd a similar sum was dispatched.

Further deliveries took place in 1810; two million dollars in all to the value of £500,000 were thus used.

As we have said already the difficulty about the copper coinage had been that the Royal Mint, with its antiquated slow methods and limited accommodation, was quite incapable of supplying anything like the quantity of coin required, but the officers would not admit any deficiency and hence occupied the position of the dog in the manger. It was decided at long last to erect a new Mint on Tower Hill. Although so advanced in years and feeling the inroads of disease, Boulton was employed to plan the buildings, and to supply the machinery, including the necessary steam engine. This he did, earning universal approbation. The plant was looked upon at the time of its erection, 1805–10 and for many years afterwards, as the most complete of its kind in the world and a model to be copied. It is of interest to know that its machinery was not superseded by new till 1881–82.[1]

Supplying machinery for mints became one of the activities of Soho Foundry, and in Boulton's lifetime Royal mints were supplied to Russia, Spain and Denmark. For the privilege of supplying these countries special permission had to be obtained, and in the case of the first-named a public "Act to enable Matthew Boulton engineer to export the Machinery necessary for erecting a Mint in the Dominions of His Imperial Majesty the Emperor of all the Russias" (39. Geo. III, 1799, cap. 96) was passed.

We have already quoted Murdock's testimony to Boulton's achievements in coining, and we will only add those of Watt, eminently qualified to judge, in the *Memoir* already drawn upon so many times:

> In short had Mr B. done nothing more in the world than what he has done in improving the coinage, his fame would have deserved to be immortalized; & if it is considered that this was done in the midst of various other important avocations, & at an enormous expense for which he could have no certainty of an adequate return we shall be at a loss whether to admire most his ingenuity, his perseverance or his munificence. He has conducted the whole more like a sovereign than a private manufacturer. The Love of fame has been to him a greater stimulus than the love of gain.

Darwin was naturally more enthusiastic than Watt and describes in poetic language the process of coining as carried

[1] *The Engineer*, January 19th, 1883.

on at the Soho Mint.[1] We make no apology for reproducing the stanzas, which are as follows:

> Now his hard hands on Mona's rifled crest,
> Bosom'd in rock, her azure ores arrest;
> With iron lips his rapid rollers seize
> The lengthening bars, in their expansion squeeze;
> Descending screws with pondrous fly-wheels wound
> The tawny plates, the new medallions round;
> Hard dyes of steel the cupreous circles clamp,
> And with quick fall his massy hammers stamp
> The Harp, the Lily and the Lion join,
> And GEORGE and BRITAIN guard the sterling coin.

In the accompanying footnotes, Darwin explains that "Mona's rifled crest" is an allusion to the "copper mines in the isle of Anglesey the property of the Earl of Uxbridge".

In regard to coining Darwin says:

> Mr. Boulton has lately constructed at Soho near Birmingham, a most magnificent apparatus for Coining which has cost him some thousand pounds; the whole machinery is moved by an improved steam-engine, which rolls the copper for half-pence finer than copper has before been rolled for the purpose of making money; it works the coupoirs or screw-presses for cutting out the circular pieces of copper; and coins both the faces and edges of the money at the same time, with such superior excellence and cheapness of workmanship, as well as with marks of such powerful machinery as must totally prevent clandestine imitation, and in consequence save many lives from the hand of the executioner; a circumstance worthy the attention of a great minister. If a civic crown was given in Rome for preserving the life of one citizen, Mr. Boulton should be covered with garlands of oak. By this machinery four boys of ten or twelve years old are capable of striking thirty thousand guineas in an hour, and the machine itself keeps an unerring account of the pieces struck.

To show the range and character of Boulton's coin and medal work, a few typical examples have been figured on Pl. X and numbered as shown below. They have been referred to already in the text but deserve a detailed description.

[1] *The Botanic Garden*, Part I, 1791, p. 27.

(1) MINT MEDAL, 1798.

Obverse—Bust of Boulton to r. on the raised rim in incised lettering: MATT· BOULTON ESQ· F·R·S·L· & ED·F·R·I· & A·S.

Reverse—In concentric circles:

M(ATTHEW) : BOULTON ERIGEA A SOHO ANGL(ETERRE) : 1788 UNE MACH(INE) : A VAPEUR P(OU)R FRAP(PER) : MONN(AIE) :

1798. IL ER(IGEA) : UNE BIEN SUPERIEURE A 8· BALANCIERS NOUVEAUX

CES CERC(LES) : & CHIF(FRES) MARQ(UENT) : LE DIAM(ÈTRE) : & NO(MBRE) : DE PIECES FRAP(PÉES) : P(AR) : MIN(UTE) :

P(AR) : 8 ENFAN(T)S SANS FATIG(UE) : DU PL(US) : PET(IT) : OU PL(US) : GR(AND) : VOLUME.

OU DE 8. DIFF(ÉRENTES) : GRAND(EURS) : ENSEMBLE. ON PEUT

AUGM(ENTER) : L'EFF(ET) : AU DEG(RÉ) NECESS(AIRE).

In the centre, head of science, Copper $1\frac{5}{8}$ in., 41 mm.

On each of the circles mentioned is a figure 400, 480, 560, 640, 720, 800, 920 which is the number of coins of the diameter of that circle that could be struck per minute. It has been observed already (p. 144) that the above legend is a description of Boulton's improvements in the art of coining and the capacity of his press. A translation of the legend has already been suggested and need not be repeated. The dies for this piece must have been sunk by Droz.

(2) TRAFALGAR MEDAL, 1805.

Obverse—Bust of Lord Nelson to l. in naval uniform. Around HORATIO VISCOUNT NELSON. K.B. DUKE OF BRONTE. &.

Reverse—Above: ENGLAND EXPECTS EVERY MAN WILL DO HIS DUTY. In the centre representation of the naval battle. Below: TRAFALGAR OCT^R. 21. 1805.

Edge—TO THE HEROES OF TRAFALGAR FROM M. BOULTON. Pewter, copper, or copper silvered. $1\frac{7}{8}$ in., 48 mm. The dies for this piece were sunk by Küchler.

One of these medals, in the appropriate metal, was presented by Boulton, with the permission of the Admiralty, to every man who took part in the engagement.

(3) TWOPENNY PIECE, GREAT BRITAIN, 1797.

Obverse—Bust of George III to r. On the raised rim in incised lettering: GEORGIUS III. D: G. REX.

Reverse—Figure of Britannia with olive branch and trident, ship in the background. On a scroll under the shield SOHO. This usually requires a reading glass for its detection. On the raised rim in incised lettering: BRITANNIA 1797. Copper, 1·6 in., 41 mm.

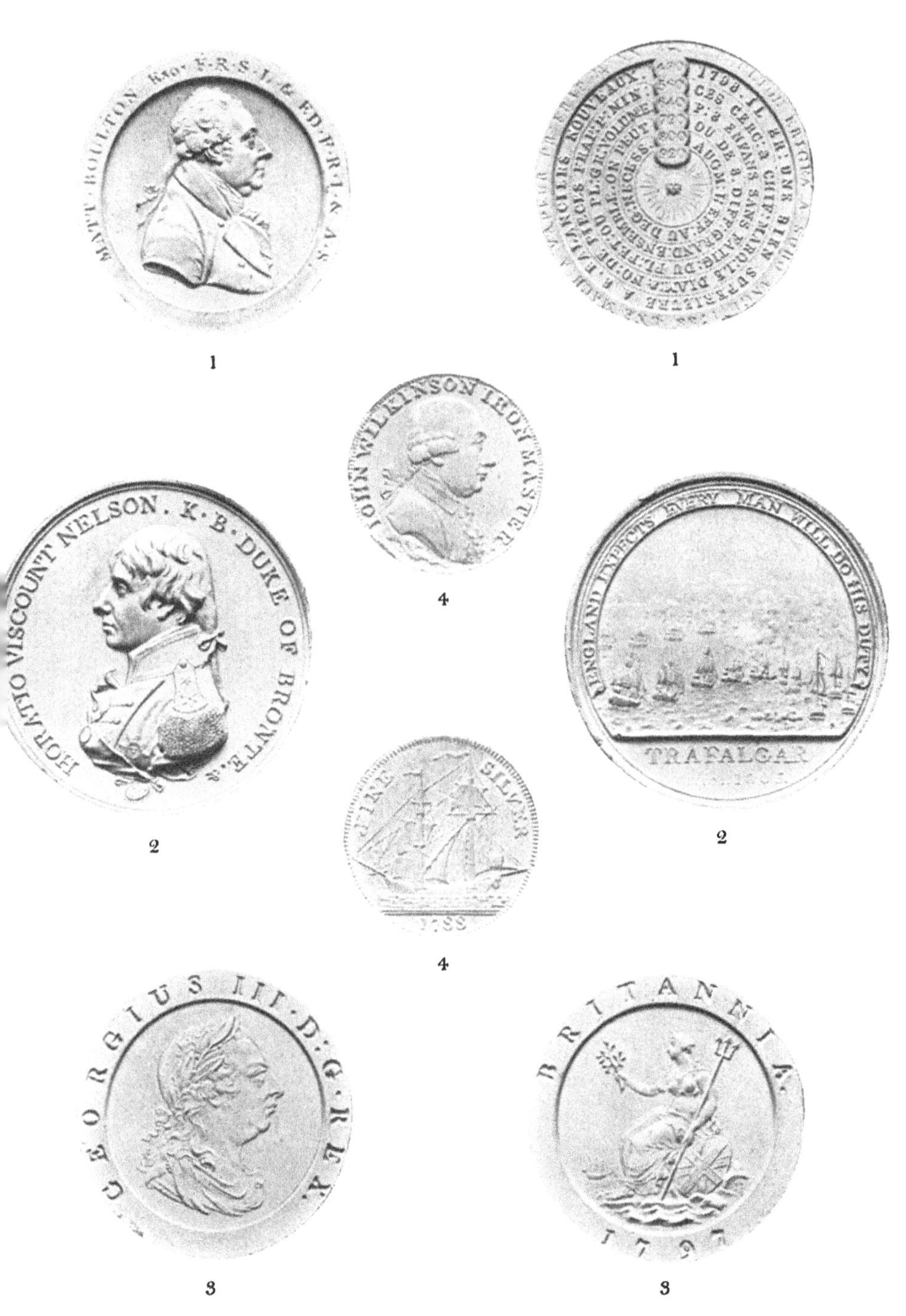

PLATE X. SOME OF BOULTON'S COINS AND MEDALS, 1788–1805

ACTUAL SIZE

Courtesy of the British Museum

(4) TOKEN, 1788.

Obverse—Bust of John Wilkinson to r. Around IOHN WILKINSON IRON MASTER.

Reverse—Brig under sail, perhaps in allusion to his iron boat. Around FINE SILVER. Below, 1788.

Edge—Bradley Bersham Willey Snedshill, i.e. the names of his ironworks where the token was payable. Silver, 1·15 in., 29 mm. Nominal value *3s. 6d.*

On p. 162 and p. 202 will be seen a medal struck at Soho in memory of Boulton for presentation to his friends. As it is in a different category to the preceding, it has not been included on Pl. X. The description is as follows:

MEMORIAL MEDAL, 1819.

Obverse—Bust of Boulton to r. Around MATTHAEVS BOVLTON. On truncation PIDGEON F.

Reverse—In a wreath of laurel INVENTAS AVT QVI VITAM EXCOLVERE PER ARTIS.

Rim—PATRIS AMICIS M·R·B CIƆIƆCCCXVIIII. Bronze, 2½ in., 63·5 mm.

The full quotation (Virgil, *Æneid* VI, 663–4) is "Inventas aut qui vitam excoluere per artis, Quique sui memores alios fecere merendo" "Men who have ennobled life by their discoveries in the arts, or who have earned by merit the remembrance of others". "To the friends of his father M(atthew) R(obinson) B(oulton) 1819."

Forrer III, 242, states that this medal was begun by Küchler, that after he left Soho, Pidgeon finished it and that the latter received £300 for the work.

A by-product of the mint was the "scissel" or perforated sheets of copper from which the coin blanks had been punched. In the ordinary way it was remelted, recast and rerolled. About this time there was much trouble in the navy and in the mercantile marine on long voyages owing to the fouling that takes place on ships' bottoms with marine growths. To overcome this, the practice was to sheathe a vessel with copper, which was expensive, or to "pay" the underwater surface with a composition that was poisonous to marine growths. Much attention was being paid to the matter at this period and it occurred to Boulton, ever alive to the problem of the hour, that copper for coinage being necessarily of high quality and malleability, would be particularly suitable for sheathing. Accordingly he wrote to the Navy Board as follows (1800, Sept. 6):[1]

Honourable Sirs, In reply to your favour of the 5th inst, the four sheets of copper you have received were intended to have been delivered by me, but I unfortunately had left town before their arrival. In consequence of the conversation that passed at your honourable Board I ordered four sheets to be rolled from my refined copper as specimens of the proposed sheathing.

		lb.	oz.
A.	Are two sheets the one is *soft* the other hard rolled to the thickness of 28 ounces per square foot to the breadth of 14 inches and length 4 feet . . .	16	6
B.	Are two sheets the one is rolled soft and the other hard to the breadth of 14 in length of 4 ft and 32 ounces per square foot	18	$11\frac{1}{2}$

I should be happy to receive your remarks respecting these sheets, and if you please to honour me with an order for as much as will sheathe one ship it shall be as smooth rolled as any you have used, and it shall be rolled to any degree of hardness or softness you may think proper, and I have not the least doubt but I can refine it so as to make it as durable as the best copper sheathing that has ever been applied to his Majesty's ships.

On November 18th the Board sent him an order, but in the meantime he had quoted for copper sheathing for the Hon. East India Co., and had been so successful that he had got the whole contract, and to complete the order he had had to take the sheets ordered for the Navy Board. Boulton was profuse in his apologies and finished up by saying that he would "hope that you will allow me an opportunity of retrieving my character, which has in this instance failed of punctuality".

The Board was annoyed but overlooked his *faux pas* and informed him that a couple of frigates would be ready by the end of January and he could "send up the quantity already demanded of him to Deptford Yard by the 30th of January at the latest". On January 8th he sent up 3300 sheets weighing

[1] Prosser, R. B., *Birmingham Inventors*, 1881, p. 122.

about 12 tons. The *Æolus* and the *Medusa*—the frigates in question—were sheathed with it on one side while ordinary sheathing was used on the other. The result was in the case of the *Medusa* that the side of the vessel sheathed with Boulton's copper was reported to be extremely foul with long grass adhering to it, whilst the other side was clean.

This result was communicated to Boulton in a letter of November 26th, 1802, to which he replied on December 24th, apologizing for the delay, "having been confined to my bed chamber from the beginning of March to this day". Boulton remarked:

I have always been of opinion that the durability of copper sheathing depended on the purity of that metal, and I assure you the whole quantity I had the honour to furnish you with was doubly refin'd for it was made from the scissel, or scrap copper remaining after cutting out the halfpence I made by order of our Government....But as all theories and reasonings are liable to prove fallacious when put in competition with actual experiment, I think it would be well if your Honorable Board were to order a regular series of experiments to be made on this subject....

He then submits a programme of suggested experiments and makes some pertinent remarks about corrosion and says: "If copper be alloy'd with iron or zinc, I know such alloys will accelerate its destruction...."

Doubtless he had here in mind the fate of Keir's alloy (see p. 101). His most acute observation is this: "I am inclined to think that the accelerated corrosion of iron bolts in the vicinity of copper will be better understood when we become more acquainted with the galvanic fluid which I am persuaded is a principal agent in that operation."

We have here quite a remarkable observation suggesting a field for scientific research and accompanied by a programme for carrying it into effect that was at least half a century before its time.

Boulton concluded with a request to have a sheet from each of the sides of the *Medusa* for examination, but alas! the copper had been melted down, the Officers of the Yard pleading that they were unaware that the ship had been coppered experimentally. There the matter ended, but what vigour and freshness of outlook for a man in his 75th year to show!

Boulton was now feeling the weight of advancing years, and he had had the first attack of the kidney trouble that later developed so seriously, but "the labour we delight in physics pain". Well could he say to his friend Garbett about this period of his career, and it serves as a fitting conclusion to this chapter:

> Of all the mechanical subjects I ever entered upon, there is none in which I ever engaged with so much ardour as that of bringing to perfection the art of coining in the reign of George III as well as of checking the injurious and fatal crime of counterfeiting.

CHAPTER VIII

SOHO FOUNDRY

Changes in manufacture—Education of partners' sons—Taken into partnership—New works—Rearing feast—Patent litigation—Prices and performances of engines—End of partnership of Boulton and Watt—Soho Insurance Society—Soho House.

WE must now hark back a few years to show the changes that were coming over the steam engine business owing to the imminent expiration of Watt's master patent.

It will be recalled that, at the very outset, even before the beginning of the partnership with Watt, Boulton had laid down in clear language his ideas as to how the manufacture of engines should be carried on (see p. 81), but in actual practice the partners fell in with the established practice of engine building, viz. of assembling the materials, including even the iron plates for the boiler, on the spot where the engine was to be erected. The partners were the more agreeable to this procedure from the fact that it was possible to obtain essential parts—cylinder, condenser, piston rod—of the necessary standard of accuracy from existing establishments, such as those of John Wilkinson and later that of Coalbrookdale. This left a few, but only a few, parts to be produced at Soho. This arrangement fell in admirably with the views of both partners: of Watt because he was absorbed in the design and development of the engine, working with his assistants at his house at Harper's Hill and only visiting the Manufactory at intervals; of Boulton, because his hands were already full with Soho manufactures, and because he was thereby saved capital expenditure on new buildings at a time when he was sorely pressed for money by the rapid expansion of the businesses in question.

With the advent of the rotative engine, a change came over the scene. Customers for such engines had no organization for collecting materials, erecting engine houses and engines or making boilers. All they wanted was to be able to order an engine and leave it to Boulton and Watt to do whatever was necessary to supply, erect and leave it in running order. The consequence was that all the complicated parts—parallel motions, valve gear, sun and planet wheels, etc., had to be supplied from Soho. The transition in practice was gradual. Evidence of it is afforded in a letter from Southern to Watt (B. and W. Coll. 1794, Feb. 15): "I should wish to hear...whether you would have anything more than the cylinder and air pump cast at Bersham and whether we shall furnish from Soho the other castings...and likewise whether the Bersham goods shall come here to be fitted or be sent from Chester to London."

Figures are available showing that by 1793 the proportion of the materials supplied from Soho had risen to about 50 per cent. of the whole. Now very little provision existed at the Manufactory for making these engine parts beyond an engine yard in the rear of the premises and at a lower level. This was provided with a smith's shop which, although Boulton refers to it as the "great smith's shop", was probably of no particular size. There was at least one lathe in this shop, and it was furnished with benches and vices so that it served also as a fitting shop. To modern eyes the whole place would have looked like nothing so much as a country blacksmith's shop, with, however, not so good an outfit of tools as such a shop has to-day.

It was fully realized both by Boulton and by Watt that, when the master patent for the separate condenser should expire in 1800 and when the other patents should expire prior to that date, they would possess a certain amount of goodwill and not much else unless they bestirred themselves. They, or probably Boulton, for Watt by this time was looking forward to retirement, decided that they would have to become engine builders in competition with all comers.

There was a further sound reason for this decision; there were the respective sons of the partners to be considered. Boulton had, as we have said, an only son, Matthew Robinson, born in 1770, who was brought up by his father with anxious care. We get a glimpse of the children Matt and Nancy on holiday in

Cornwall in 1784—they were motherless by this time—spending the time with their father who was there on business. Boulton wrote to Watt:

I shall be happier during the remainder of my residence here than in the former part of it; for I am ill calculated to live alone in an enemy's country, and to contest lawsuits. Besides, the only source of happiness I look for in my future life is in my children. Matt behaves extremely well, is active and goodhumoured; and my daughter too, has I think good dispositions and sentiments which I shall cherish.

We hear of Matt helping his father in chemical experiments and playing with toy balloons filled with hydrogen. As one would expect the father instilled into the son the duty of cultivating good manners. He studied under a tutor to whom Boulton wrote: "let him [Matt] not neglect the present but apply himself so as to become well grounded in Grammar and Latin...he is capable, but not of close application, to which he must be inured, as no proficiency of any kind can be acquired without it." Boulton took his son over to Paris towards the end of 1786—he was only sixteen years old be it remembered—and placed him under the care of a competent tutor. His father wrote to him regularly, telling him to keep out of bad company and keep down his personal expenses, for he was spending more than his father thought good for him. Always the letters were yearning (1787, Dec. 19): "There is nothing on earth I so much wish as to make you a *man* a good man a useful man and consequently a happy man." No wonder that Matt applied himself assiduously and became competent in French, German and chemistry. Writing to Mathews (1788, Aug. 25) Boulton says: "Matt is a tolerable good chemist...I shall be glad when the time arrives for him to assist me in the business."

In the summer of 1788 Matt paid a holiday visit to Soho, and Mary Anne Galton[1] describes her "astonishment at his full dress in the highest adornment of Parisian fashion", and how all the members of the Lunar Society, to a meeting of which he was introduced by his father, hung on his descriptions of the

[1] *Life of Mary Anne SchimmelPenninck* (her married name), 1858, p. 125.

discontent that presaged the French Revolution. Matt went back for another year but returned when the revolution broke out.

With regard to Nancy it may be said that her father wrote to her most affectionately whenever he was away from home. As she grew up she appears to have taken over, with her cousin and companion, Miss Mynd, the charge of Soho House. But to the great grief of her father she developed some complaint which crippled her; her father called her his "dear helpless daughter" and, in spite of all that physicians could do for her, she never recovered.

The education of young Watt was equally well cared for by his father. He was sent to Wilkinson's Works at Bersham, where he worked in the carpenter's shop for three hours a day and studied bookkeeping, geometry and algebra. He was, by Boulton's influence, placed in the counting house of Taylor and Maxwell, fustian makers, of Manchester: the little incident of his getting into debt and Boulton's rescue of him will be found told elsewhere (p. 194). After two years there, influenced by Thomas Cooper and the stirring events in France, he went as a delegate for the Constitutional Society of Manchester to Paris in 1792; for this he was denounced by Burke in the House of Commons. Young Watt fell under suspicion in Paris of being a spy, and this led to his flight to Italy and by roundabout ways home in 1794. He was a marked man, but as for over two years he had had no communication with any of the revolutionary societies, the storm blew over.

There was another son of Watt by his second wife, Gregory, born in 1777, but he was still at college. He grew up a lad of great promise, but to the great grief of the parents he developed consumption and died in 1803 in his twenty-seventh year without having left any imprint on the story we are telling.

To resume our narrative. Each partner now had a son ready to succeed him: young men of intelligence and character and energy; well educated for business, linguists, and acquainted with natural science; and of a calibre much above the average of their contemporaries in the mercantile world. They now fulfilled parental hopes by entering the firm.

They were allowed to try their wings by taking over the copying-press business, the patent of which was due to expire in 1794. They took up the manufacture vigorously, and James Watt, junior, designed a self-contained portable press which was fitted to hold all the necessary materials for copying, and opened out to form a writing desk. These desks were made in large numbers both in foolscap and quarto sizes for home and export. A few have been preserved, notably in the Boulton and Watt Collection and in the Science Museum, South Kensington; occasionally one is to be met with for sale.

In October, 1794, a new firm under the style of Boulton, Watt and Sons was formed, comprising, besides the two elders who found the money, Matthew Robinson, James junior and his brother Gregory. They now decided to take the long contemplated step of erecting an entirely new works with every convenience that experience could suggest. Here was a magnificent field for the energies of the junior partners. Boulton was enthusiastic, but Watt seems to have kept in the background except in a financial capacity.

The selection of the site showed good judgment. The Manufactory had been established before the canal era; now the partners decided to ensure having water communication, especially in view of the heavy and bulky nature of the intended products as compared with those of the old works. Consequently a site near the Birmingham and Wolverhampton Canal in Smethwick, not much more than a mile distant from the Manufactory, was selected. The area was 18½ acres or thereabouts. Negotiations were quickly concluded and the purchase was completed by James Watt on August 27th, 1795.

The layout of the works received close attention; it was "well digested and settled previous to laying the first stone; the whole is thereby rendered more complete than such works as generally arise gradually from disjointed ideas".[1] A cutting to the canal and a wet dock to accommodate four barges were made. The buildings erected comprised a smithy, a foundry with an

[1] Shaw, *Staffordshire*, 1798, II, p. 119.

air furnace, boring mill, turning, fitting, erecting and carpenters' (i.e. pattern-makers') shops.

Work was prosecuted with all speed, and there was a good reason for haste. John Wilkinson and his brother who were in partnership at Bersham had not been agreeing very well. What exactly was the point at issue is not known, but it appears that William who had been mostly abroad on business for the firm thought he was not receiving his due share of the profits. John Wilkinson was an egotistical, dominating and unscrupulous person, ready to crush anyone who crossed his path, whether a relative or not. The upshot of the dispute was that he determined to close Bersham Ironworks, the foundry where Boulton and Watt had for so many years got their best cylinders. This was a serious matter and was the last straw; the partners determined to be no longer dependent on outsiders.

Soho Foundry therefore had to have a boring mill and it had to be of the latest design. To Peter Ewart, a millwright of Manchester, later to become well known as the chief mechanical engineer of Portsmouth Dockyard, was entrusted the task of designing the mill; the design was a vertical instead of the usual horizontal design and was later successfully carried out by William Murdock.

The new foundry was completed in the short space of three months by the end of 1795, so energetic had been the new partners, and on January 30th, 1796, the rearing feast, quite a memorable affair, took place. We cannot do better than quote the account of it that appeared in the local press.[1]

SOHO FOUNDRY

On Saturday last the Rearing Feast of the new Foundry lately built by Messrs BOULTON, WATT & SONS at Smethwick, was given to the engine smiths, and all the other workmen employed in the erection.

Two fat sheep (the first fruits of the newly cultivated land at Soho) were sacrificed at the Altar of Vulcan and eaten by the Cyclops in the Great Hall of the Temple which is 46 wide and 100 feet long. These two great dishes were garnished with rumps and rounds of

[1] *Aris's Birmingham Gazette*, February 1st, 1796.

beef, legs of veal, and gammons of bacon, with innumerable meat pieces and plumb puddings, accompanied with a good band of Martial Music. When dinner was over, the Founder of Soho entered, and consecrated this new branch of it, by sprinkling the walls with wine, and then, in the name of *Vulcan*, and all the Gods and Goddesses of *Fire* and *Water*, pronounced the name of it SOHO FOUNDRY, and all the people cried *Amen*. A benediction was then pronounced by him upon the undertaking, and a thanksgiving offered for the protection and preservation of the lives and limbs of the workmen during the erection. These ceremonies being ended, six cannon were discharged, and the Band of Music struck up *God save the King* which was sung in full chorus by two hundred loyal subjects. After this many toasts were given suitable to the occasion, by the President of the Feast (Mr M. ROBINSON BOULTON) which was conducted by him with great spirit and hilarity; each toast was accompanied with three joyous huzzas and a discharge of cannon. A Ball with tea, was given in the evening to *Venus* and the *Graces* which ended about ten o'clock, when the concluding guns were fired, and all departed in good humour.

The Address of Mr BOULTON Sen. upon entering the Foundry (after making an excuse to the Company for not dining with them on account of ill health) was conceived in the following *terms*:[1]

"I could not deny myself the satisfaction of wishing you a happy & joyous day & of expressing my regard for all good honest & Faithfull workmen whom I have always employed.

"I now come as the Father of Soho to Consecrate this place as one of its Branches, I also come to give it a Name & my Benediction.

"I will therefore proceed to purify the walls of it by the sprinkling of wine and in the name of Vulcan & all the Gods & Goddesses of Fire & Water, I pronounce the name of it *Soho Foundry*—May that name endure for ever & let all the people say amen.

"This Temple now having a name I will propose that every man shall fill his pitcher & drink success to it. I will now call your serious attention whilest I give my Benediction to Soho Foundry.

"May this Establishment be ever prosperous, may it give birth to many usefull Arts & Inventions. May it prove beneficial to Mankind & yield comfort & happiness to all who may be employ'd [in] it.

"As the Smith cannot do without his Striker so neither can the Master do without his Workmen. Let each perform his part well

[1] The text of the speech is taken from the holograph original and not from the newspaper. The versions are practically identical.

and do their Duty in that state to which it has pleased God to call them & this they will find to be the true ground of Equality.

"One serious word more & then I have done. I cannot let pass this Day of Feastivity without observing that the piles of Building have been erected in a short time without the Loss of one Life or any material accident. Therefore let us offer up our gratefull thanks to the divine preserver of all things without whose permission not a sparrow falls to the ground. Let us chaunt in our Hearts Hallalujah for these divine Blessings and with our voices let us like Loyal subjects sing God save ye King."

It is not difficult to imagine the happy scene, with the now patriarchal Boulton moving among his workmen like a king among his subjects. It is a pleasure to be able to say that his wish that the establishment should prosper and that it should give rise to useful inventions, has been fulfilled. It has had its ups and downs but it has never been more prosperous than at the present day. Soho Foundry is now engaged in the production of weighing machinery by the firm of Messrs W. and T. Avery, Ltd., proprietors of the firm of James Watt and Co. In this way the name lives on. We wish the same could be said of the Manufactory but it was swept away in 1848.

The new venture was, as we have said, financed by Watt and by Boulton. The latter writing to Capt. Apsley (1796, Jan. 26) says: "I and my partner James Watt have laid out £10,000 in erecting a Foundry on the bank of the Birmingham Canal for the purpose of casting everything relating to our steam engines."

The Foundry was at the outset quite a small concern but even so there was difficulty in staffing it. By the closing of Bersham Works, several of Wilkinson's men were out of work and some of these were secured. Abraham Storey, head foundryman at Bersham, and John Kendrick, from the same place, were engaged. John Southern was sent up and down the country to engage workmen. William Murdock, who was in Cornwall and had received favourable offers from the adventurers to remain there, was proof against their blandishment and returned to the fold at Soho somewhere about 1798. As in Cornwall, he proved a tower of strength to the firm; apart from his own valuable inventions in connection with the steam engine—such as the long D slide valve and the eccentric—he built up the workshop practice by his own example.

Looking after the drawing office was John Southern, a man of some mathematical ability, able to calculate and design the many new applications of the engine that were being continually called for. With the energy and aptitude of the junior partners, this was a strong, nay an ideal, combination and it is not to be wondered at that Soho Foundry throve.

It has to be admitted that in the beginning the work turned out at the Foundry was inferior to that of some of their competitors, for example, Matthew Murray of Leeds. This defect was slowly remedied, and in time Soho reached the front rank of engine-building establishments. An apprenticeship there was a passport to almost any mechanical engineering job in the country.

A word or two should be said about the partnerships. The new firm of Boulton, Watt and Sons formed in 1794 had as partners Boulton and his son, Watt and his two sons; the proportions held by the two families were the same as in the original partnership. The original firm continued in existence, but all new work coming forward was handed over to the new firm, "the future manufactory [i.e. Soho Foundry] and profits of steam engines made by them". At the end of the year 1800 Watt gave up his small share to his two sons and the firm was equally divided as concerned both pumping and rotative engines between the two families; the firm became Boulton, Watt and Co., Soho. On the death of Gregory Watt his share reverted to his brother, and on Boulton's death his share was bequeathed to his son, so that in 1809 James Watt, junior, and Matthew Robinson Boulton were the sole partners, each holding one-half.

Now as to Soho Foundry. James Watt held no share; he merely paid for the estate, and this he handed over in 1801 for the price he had paid for it. Boulton with his son and the two Watts were the partners, so that individually and by families they held equal shares. The firm was "Boulton, Watt and Co.,

Soho Foundry". The same thing happened as with the previous firm, so that in 1809 the position was James Watt, junior, and Matthew Robinson Boulton, sole partners with equal shares.

A word must be said about the organization and running of the Foundry. James Watt, junior, seems to have taken this particularly under his wing. Hundreds of documents survive to show how he went into costs of each of the shops, the labour force required for practically every operation, the time required, the cost of it and, if performed by a tool like the boring mill, the feeds and speeds likewise. In no other establishment then in existence did anything comparable exist, indeed the organization will bear comparison with that of modern establishments. Boulton did not, so far as we can gather, concern himself in any of this work.

The Foundry as one would expect turned out to be most profitable. The former delays in obtaining castings and forgings were obviated, but apart from that the firm, owing to their reputation, were able to charge and to obtain higher prices than their competitors. The capital cost of the Foundry, which had reached £27,000, was paid for out of profits by 1804. Further extensions were called for, and in 1816 the costs of these extensions amounting to roughly £20,000 were likewise paid for out of profits. We cannot pursue the subject further as it is outside the scope of the present volume,[1] but the firm was a positive gold mine and the junior partners reaped what the older ones had sown.

The Foundry is shown in the perspective view on Pl. XI taken from a water-colour sketch of date about 1820. The wet dock in communication with the canal can be seen, but we have been unable to determine the allocation of the various buildings. Practically none of them now remains except the entrance gateway (see Pl. XII), the position of which can be identified without much difficulty. In this gate-house is a clock, dated 1802 and still keeping good time. On the wall below the clock will be observed a tablet which has this inscription: "Boulton

[1] But cf. Roll, E., *An Early Experiment in Industrial Organization*, 1930, Part II.

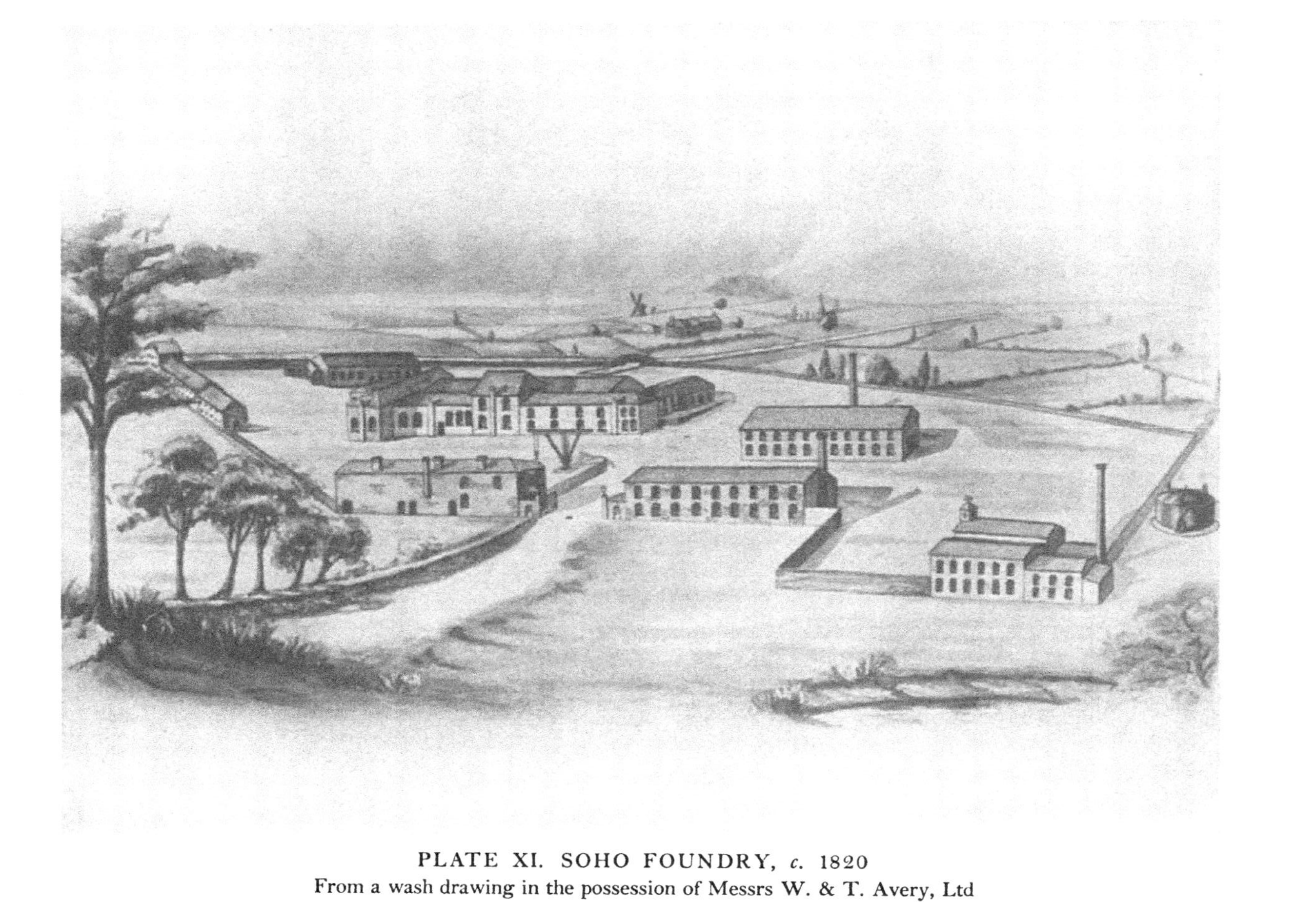

PLATE XI. SOHO FOUNDRY, *c.* 1820

From a wash drawing in the possession of Messrs W. & T. Avery, Ltd

PLATE XII. ORIGINAL ENTRANCE TO SOHO FOUNDRY

Courtesy of Messrs W. & T. Avery, Ltd

& Watt installed this clock in its present position in the year 1802. Beneath this clock stood the entrance gates of Soho Foundry." The place where the gatepost stood beside the door can be seen. It may sound sentimental but we are pleased to say that when the adjoining buildings were demolished this gatehouse was left standing and is now incorporated in the new buildings on the site.

Before leaving the Foundry and all its works we ought to mention Montgolfier's hydraulic ram for raising water, because the question of taking up its manufacture as one of the firm's products seems to have been seriously entertained. Boulton took out a patent for the ram under the title "hydraulic engine" in 1798 Jan. 30 (No. 2207). He does not mention in the specification that it was a communication from the inventor, Joseph Michel Montgolfier (1740–1810), but the existing correspondence shows such to have been the case. This omission has led to misunderstanding and even to accusations that Boulton pirated Montgolfier's ideas. A memorandum exists of calculations by John Southern of the effects of the hydraulic ram, showing that it was carefully considered, but the decision evidently was against it. This appears somewhat strange as it would have been a suitable machine to manufacture in the Foundry. There were bigger fish to fry and perhaps that is why it was not taken up. All we can say is that it was not till 1824, owing to the exertions of James Easton, that the hydraulic ram was brought into general use either in this country or abroad.

We have mentioned repeatedly the difficulty the partners had with infringers and pirates of the condenser patent. Jonathan Hornblower, whose compound engine of 1781 embodied the condenser, had not been a serious danger as his engine was only erected on a limited number of mines, and when he tried to get an extension of his patent from Parliament in 1795 he was successfully lobbied out of it by the exertions of the friends of the partners. Other infringers had given way on the threat of legal proceedings, but there were two whose opposition could not be overcome, Jabez Hornblower and Edward Bull.

Jabez was the brother of Jonathan (there were eight of these brothers, each with a biblical name beginning with "J"), and had been employed by the firm in Cornwall in 1779, but he was of a surly disposition and his employment was not continued. After a varied career he came to London in 1790 and set up as an engine maker. In 1795 he entered into partnership with John Maberley, who had bought a patent of Isaac Mainwaring for a double-cylinder engine.

So far so good, but it could not be made without the separate condenser and that constituted an infringement of Watt's patent. Injunctions were served but all to no effect, and to cut a long story short the case was brought to the Court of Common Pleas on December 16th, 1796, before Lord Chief Justice Eyre. The verdict was for the plaintiff. Boulton was in court and wrote to his daughter the following day thus:

> I am in perfect health though much hurry'd, but cannot withold from you & Miss Alston [Miss Boulton's housekeeper or companion] a participation of the Joy we all feel from the result of our Tryal which began at ten yesterday morn[g] & though we did not call half of our Witnesses it did not end till half past 8 at night when the Jury gave us a full Verdict with cost of Suit.

There was much rejoicing at Soho when the result became known. The defendant, however, brought further proceedings in the same Court on a writ of error affirming that Watt's patent itself was invalid—on the ground that the description was not enough to enable a qualified person to construct the engine. It must be admitted that the patent, as we have indicated already, was rather shaky on this point. Eventually the original judgment was affirmed.

Parallel with this case, another one, that with the recalcitrant Edward Bull, had dragged on a weary course through the Law Courts. He was really a man of straw, but he was backed up by the Cornish adventurers who were using him as a cat's-paw. He too had been employed originally by the firm in 1781 and had been sent into Cornwall when the very busy time was on. He had been seduced from the firm's interests and became attached to those of the adventurers. He invented in 1792 an inverted engine, known as the Bull engine, which was neither better nor worse than one that Watt had invented as early as 1766. The trouble with Bull's engine and indeed with everyone else's design was that without the addition of Watt's condenser it was of no particular value. An injunction was served but, as is so common, no notice was taken of it and no other course remained open except to take Bull to court. The case came on for hearing in the Court of Common Pleas on June 22nd, 1793. The verdict was for the plaintiffs, subject to the opinion of the Court as to the validity of the patent. On May 16th, 1795, this special case came on for hearing but was inconclusive as the judges were equally divided. This brought on another trial and after interminable delays, because the case Boulton and Watt *v.* Hornblower was also involved, the Judges in the Court of King's Bench affirmed the validity of the patent in January, 1799.

Broadly speaking it was a triumph of common sense over legalism and therefore a credit to the Courts. However defective Watt's patent might be, there was no defect in the nature and importance of his invention. Quite apart from this, much can be said as to whether Parliament in its wisdom ought to have given Watt such a long extension of his patent as it did; the development of the steam engine was certainly held back by it many years. Damages of about £6000 were awarded to Boulton and Watt, but they recovered only about £2000. Their own legal expenses are said to have amounted to £10,000 so that, like most actions at law, it was a ruinous one. However, the decision of the Courts was of great importance to Boulton and Watt because the firm was now enabled to come down on the swarm of adventurers and others who had refused to pay their premiums, awaiting the result of the trial. Upon the task of making them pay up, the junior partners entered with the utmost zest. Boulton himself seems to have taken the part of spectator but he benefited in pocket greatly.

Some idea of what were the amounts involved in the dealings of Boulton and Watt with the Cornish mine adventurers is afforded by a paper drawn up by Mr Thomas

Wilson, their representative and agent in Cornwall, dated 1799, Feb. 22.

Total amount of fuel saved	£803,869. 19. 8
One third part due for premium	267,956. 13. 0
Amount rec[d] by Boulton & Watt	105,904. 9. 5
Short	162,052. 3. 7

Only a small amount of this shortage was recovered in spite of all the efforts of the junior partners. It is safe to say that no patent up to that time had ever brought in so large a sum of money.

It is a matter of considerable interest to know what was the total number of engines built during the partnership, 1775–1800, for what purpose or duty they were built, and how they were distributed geographically. Much information under these heads is furnished by the map on the opposite page. The coalfields, the metalliferous areas and the textile districts as they existed in 1800 are shown. The number of engines, divided according to type: sun and planet, crank, pumping and blowing engines, in each county is marked by symbols. It will be observed that the reciprocating engines number 188 and the rotative 308, showing how right Boulton was when he urged Watt to devote his attention to the latter type as affording their largest market. Of the rotative engines by far the largest number were for the textile industry.

If we average the horse-power of each engine as 15, which is fairly near the mark, we get a total horse-power of 7440—in round figures 7500 H.P.—built in twenty-five years. From this trickle has grown the mighty stream of mechanical power we see to-day, when a single unit for an electric power station may easily be nine or ten times the capacity added together of all the engines built by the firm in a quarter of a century.

Another question that may be asked is, what were the prices at which Boulton and Watt sold their rotative engines at the end of the partnership when the premium or royalty system of charging for engines had disappeared? This information is furnished by a letter signed on behalf of the firm by Gregory

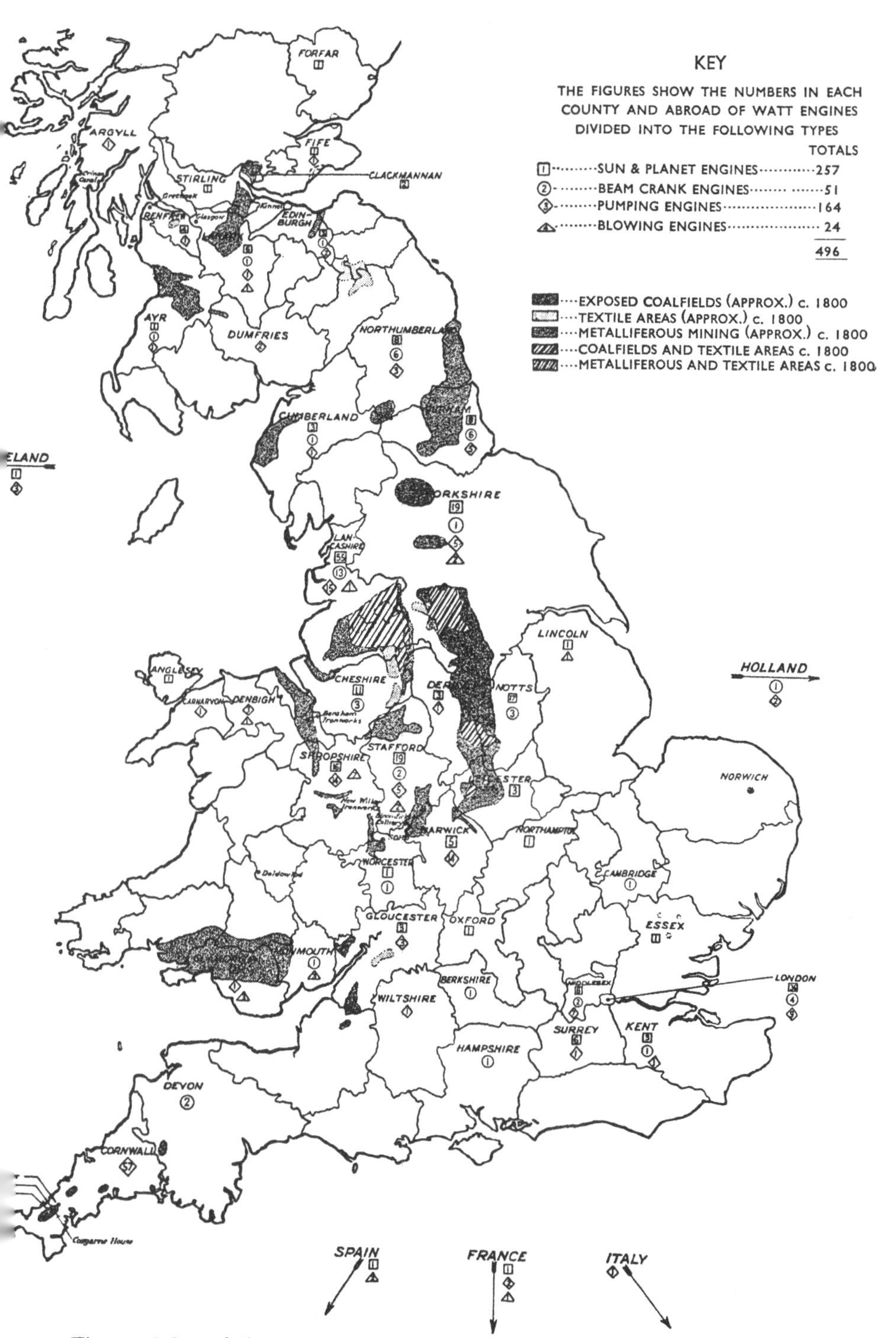

Fig. 6. Map of distribution of Boulton and Watt Engines, 1775–1800.
Courtesy of the Science Museum.

Watt, addressed to Simon Goodrich, Mechanist to the Navy Board.[1]

A four Horse engine including one Iron Boiler delivered here, and payable in three months will amount to £327

10 Horse	Do	Do	to	491
12 Do	Do	Do		525
20 Do	Do	Do		750
30 Do	Do	Do		1050
40 Do	Do	Do		1440
50 Do	Do	Do		1727

Any of the smaller engines could be delivered in three months from the time of receiving the order and any of the larger sizes would not require more than between 3 and four months.

A note on the back of the letter in another handwriting gives the further information:

	Inches
Diar of a 6 Horse Engine	13½
— 10 Do —	17½
— 8 Do —	16
— 20 Do —	24

B & W allow 1 Bushel of coals per hour for every 10 Horses.

Incidentally these prices show what profits were to be made at this date.

Another aspect of the steam engine is its capacity for doing work. Boulton states it succinctly towards the close of the partnership in a letter to James Watt, junior, thus (B. and W. Coll. 1796, Nov. 28):

One bushel (84 lbs) of Newcastle or Swansey coal will

(1) raise 30 million lbs. of water 1 foot high
(2) Grind & dress 10, or 11, or 12 bushels of wheat according to the state of it
(3) Turn 1000 or more cotton spinning spindles per hour
(4) Roll and slit 4 cwt of bar iron into small nailors' rods
(5) Do as much work per hour as 10 horses.

[1] Science Museum, Goodrich Papers, 1800, Apr. 26.

The end of the original partnership of Boulton and Watt had now come. The greatest work of their lives was done, the patent had been vindicated and they were in flourishing circumstances. Boulton was now in his seventy-second and Watt in his sixty-fourth year. To the latter it was a happy release and he hardly set foot in the works again. To Boulton it was but another milestone on his life's journey; he had still his favourite coining machinery with which to busy himself.

We must now revert for a short time to Soho Manufactory. It had experienced many changes in the way of one extension after another. We are to imagine the manufacture of toys, Sheffield plate and silver plate as going on steadily in a hum-drum fashion. The number of work-people had increased to nearly 1000; "Old Bess" continued to pump water from Soho Pool to the mill pool above, to help supply the waterwheel; her power had been supplemented by that of the "Lap" engine. For the Mint an entirely new building had been erected and power was furnished by a separate engine.

Little occurred of sufficient importance to call for remark except the establishment in 1792 of the "Insurance Society belonging to the Soho Manufactory", one of the first, if not the first, mutual friendly society in this country established by a manufacturer for the benefit of his work-people. What Boulton's intention in founding it was we do not learn, but knowing his character we can safely say it was paternalism. He believed himself to be, as Boswell had said, the father of his tribe, and he felt it his duty to live up to the position. The personal touch between master and workman always subsisted at Soho, as witness Boulton's custom of giving Christmas presents to all employes in the shape of money, clothing, books or other articles. This is not the place to discuss the genesis of friendly societies and the view sometimes held that they were direct descendants of the medieval gilds; it need only be said that the urge of human beings to associate for mutual aid is always with us and only needs to be directed into an appropriate channel; it was this need and this situation that Boulton visualized.

The only earlier instance of an organization approaching the one at Soho that we know of is that established by Sir Ambrose and John Crowley at their works at Winlaton, near Newcastle-upon-Tyne. This, however, was more in the nature of a compulsory contribution imposed from above on the work-people, for insurance against sickness, old age and death, than of a friendly society.

The Soho Society embraced much more than this. Elaborate rules were drawn up and printed; had space permitted we should have liked to have reproduced them.[1] We reproduce, however, the engraving that appears at the head of the Rules (see Pl. XIII). It is of interest to know that the copper-plate still exists and is preserved in the Boulton and Watt Collection. This engraving is quite a delightful piece of eighteenth-century allegorical composition in which we cannot help believing Boulton had a hand, and we must not deny the reader the "Explanation of the Plate" which is as follows:

A Member of this Society with his Arm in a Sling, is seated on a Cube, which is an Emblem of Stability, as the Dog at his Feet is of Fidelity; he is attended by Art, Prudence, and Industry, the latter of whom raiseth him with one Hand, and with the other sheweth him Plenty, expressed by the Cornucopia lying at the Feet of Commerce, from whence it flows. Art resteth on a Table of the Mechanic Powers, and looks up to Minerva, Goddess of Arts and Wisdom, who, descending in the Clouds, directs to the SOHO MANUFACTORY, near which are little Boys busy in designing &c., which shew that an early Application to the Study of Arts, is an effectual Means to improve them; the Flowers that are strewed over the Bee-Hive, represent the Sweets that Industry is ever crowned with.

The Society was self-administered except for an over-ruling power held by Boulton. The affairs of the Society were administered by two governing bodies, both of which were required to meet once a week. The first of these bodies consisted of six members who served for three months and on retirement chose another six to take their places; later three members only retired each quarter and the whole committee elected their

[1] This is the less necessary because it has been done already by Roll, *Industrial Organization*, 1930, frontispiece. The account that follows is taken from his book, pp. 228–36.

PLATE XIII. ENGRAVING AT THE HEAD OF THE SOHO INSURANCE SOCIETY'S RULES, 1792

Courtesy of the Public Libraries Committee, Birmingham

successors. The second body consisted of six elders appointed each quarter by Boulton; they could be removed by him in case of dereliction of duty. These elders supervised the conduct of the committee and acted as arbitrators in case of dispute. Boulton retained ultimate control over the Society's affairs, coupled with the power of making new rules and abolishing existing ones. Originally the Society included all persons employed in the Manufactory whose earnings were between 2*s*. 6*d*. and 20*s*. per week. In 1796 the employes of Soho Foundry were admitted on equal terms.

A new member paid an entrance fee of 1*s*. if a man, 8*d*. if a boy between 14 and 18, and 6*d*. if under 14. He received a copy of the rules. The scale of contributions and of benefits varied according to the earnings, per week, as shown by this table:

Wages	Contribution	Benefit
2*s*. 6*d*.	½*d*.	2*s*.
5*s*. 0*d*.	1*d*.	4*s*.
7*s*. 6*d*.	1½*d*.	6*s*.
10*s*. 0*d*.	2*d*.	8*s*.
12*s*. 6*d*.	2½*d*.	10*s*.
15*s*. 0*d*.	3*d*.	12*s*.
17*s*. 6*d*.	3½*d*.	14*s*.
20*s*. 0*d*.	4*d*.	16*s*.

As time went on the scale of benefits was found to be too liberal, and eventually it had to be halved. We do not learn that there was any employer's contribution. There was, in addition, a funeral benefit scheme in return for weekly contributions.

In the edition of the Rules printed in 1804, there is a mention of "Club Rooms of Soho and Soho Foundry" which must have been a big step in advance. There are very many other provisions among the rules that are of great interest but they cannot be commented on here; the reader is referred to the work quoted for such details.

The Society concerned itself with other matters than insurance as witness Article XXIV which recites:

As it is for the health, interest, and credit of the men, as well as masters, to keep this MANUFACTORY clean and decent, it shall be deemed a forfeit of one shilling to the box, for any one found guilty of any indecencies, or keeping dirty shops, which indecency &c. shall be judged by the committee, and the forfeit made, either more or less than one shilling, according to the greatness of the indecency &c.

The question will be asked what effect had the Society on the work-people and to this we regret to say we find no answer in the firm's record. We must fall back therefore on a contemporary statement:[1] "The rules of this manufactory have certainly been productive of the most laudable and salutary effects, and besides the great attention paid to cleanliness and wholesome air, etc., this manufactory has always been distinguished for its order and good behaviour and particularly during the great riots at Birmingham" (i.e. the Church and King riots in 1791).

How long the Insurance Society lasted we do not know, but we believe it did not long survive the death of its founder. The old order changed with the advent of the younger partners to complete control, and their view, shared by the newer capitalists of the time, that labour was a mere commodity, like steam power, began to take the place of the paternalism that subsisted between master and man in the good old days of Matthew Boulton.

Let us conjure up these palmy days of Soho by listening for a moment to the paean of praise from the lips of the local poet who, moved by the scene, burst into song thus:[2]

> On yonder gentle slope, which shrubs adorn,
> Where grew, of late, "rank weeds", gorse, ling, and thorn,
> Now pendant woods, and shady groves are seen
> And nature there assumes a nobler mien.
> There verdant lawns, cool grots, and peaceful bow'rs
> Luxuriant, now, are strew'd with sweetest flow'rs
> Reflected by the lake, which spreads below,
> All Nature smiles around—there stands SOHO!
> Soho!—where GENIUS and the ARTS preside,
> EUROPA's wonder and BRITANNIA's pride;
> Thy matchless works have raised Old England's fame,
> And future ages will record thy name;

[1] Shaw, *Staffordshire*, 1798, II, p. 121.

[2] Bisset, James, *A Poetic Survey round Birmingham*, (1802), p. 12.

Each rival Nation shall to thee resign
The PALM of TASTE, and own—'tis justly thine;
While COMMERCE shall to thee an altar raise,
And infant Genius learn to lisp thy praise,
Whilst Art and Science reign, they'll still proclaim
THINE! ever blended, with a BOULTON's name.

Bisset's poetic effusions, as one can readily believe, did not bring much grist to the mill, and although he published a *Directory* and other small matters he got into deep water. It is gratifying to learn that Boulton, with his never-failing generosity, helped this lame dog over the stile.

While on the subject of the Manufactory, we cannot resist telling the story of the attempted burglary that took place there at Christmas, 1800. Boulton received word from his watchman on the premises that some men had tried to bribe him to let them through the outer door. It appeared that they were men who had previously been employed there, hence they knew their way about. They had got possession of false keys for the door of the counting house. This, not the Mint, was their objective, because not only were the week's wages there but also extra booty in the shape of cash, etc., for the Christmas presents that Boulton was in the habit of distributing.

Boulton, his son, young Watt and quite a number of the staff, with blunderbusses and other weapons, were stationed in different parts of the building. Sure enough the burglars with dark lanterns came and were let in by Boulton's instructions. They tried their keys but failed to open the door of the counting house, so retired without the watchers making any move. Boulton, writing as was his wont "to my dear Nancy", his daughter, who was staying in London, sent an account of the attack. He said: "The best news I can send you is that we are all alive; but I have lost my voice and found a troublesome cough by the agreeable employment of thief watching."

Next night—the night of December 23rd—the burglars came again, tried the door again unsuccessfully and this time they forced it. They were allowed to seize the booty and were just making off with it when Boulton gave the signal to seize them.

A quantity of tow soaked in turpentine was set fire to to light up the scene, four of the burglars were seized but the fifth, a man named William Foulds, in the scuffle and darkness got away. The four men were secured in custody till the morning. Boulton got to bed about 2 a.m. but was aroused again by a noise outside the house and fearing the advent of more burglars, he crept out stealthily and fired a gun in the direction of the sound. Let Boulton tell the sequel in another letter to his daughter (1800, Dec. 27): "In the morning I found our gray pony had been graseing under the windows & had been more frightened than I was. We have examined [her] but find she is not hurt." What a laugh they must have had!

Foulds, the man who escaped, was arrested and all five men were committed to Stafford Assizes. Foulds was sentenced to a long term of imprisonment but the rest got off. Sir Walter Scott was told about this exploit and said: "I like Boulton; he is a brave man." Most persons will agree that it was plucky conduct for a septuagenarian.

Any mention of Soho brings to one's mind that part and parcel of it—Soho House. A dwelling-house, as we have seen, had been built at Soho at the same time as the Manufactory, but on higher ground commanding an extensive view. When Boulton went to live there, which we assume to have been about 1767 soon after his second marriage, it is probable that he made considerable additions to it in order to accommodate the crowds of visitors that he entertained there. When more prosperous days arrived it was natural that he should desire to have a grander house, but he did not seek to gratify his wishes, as did Watt, by building on an entirely new site. In 1789 Boulton commissioned Samuel Wyatt, the architect who has been mentioned already, and to judge by what we can see today, for Soho House is still standing (see Pl. XIV), he must have practically rebuilt the place. The elevation is dignified and the building most substantial. Boulton seemed determined to have a weatherproof, warm and comfortable house, for the exterior is covered with slates—they look like ashlar masonry in the illustration—and within he had a hot-water heating

PLATE XIV. SOHO HOUSE, PRESENT DAY

Courtesy of the Assay Office, Birmingham

installation. In several of the rooms there are delightful Adam fireplaces, while the plasterwork is in keeping. The doors and architraves are of mahogany, and there are real marble pillars in the hall. A point of some interest is that the French window sashes are of metal—without much doubt Keir's metal, which we have described elsewhere.[1] On the roof there is a lead flat where Boulton is said to have had an astronomical telescope. In the floor of the kitchen is a stone slab which covers the well which formerly supplied the house with water, and is in fact the well of the original warrener's cottage that first stood on the site. The cellarage is extensive, including the inevitable wine cellar, with slate shelves, for Boulton used to get in a pipe of wine at a time. Altogether the house and its appointments give a feeling of dignity combined with comfort and elegance—the home of an English country gentleman of the eighteenth century.[2]

The surroundings of the house, now alas shorn of their dignity by building encroachments, were in keeping. As Watt said he turned "a barren heath into a delightful garden" by planting trees and treating the millpool and the lower or Soho Pool as ornamental water. Matured by the lapse of time, Soho made a restful oasis that long has diverted on either hand the onward march of bricks and mortar from Birmingham. We trust that the house may long be preserved, for it is something to be able to stand in the room where the Lunar Society met and try to recapture the atmosphere of that bygone age. The house is now a residential hotel, and it is, we are glad to say, in sympathetic hands.

[1] See p. 103.

[2] The house was considered worthy of inclusion in the Country House series of *Country Life*. The description with many illustrations appeared in that newspaper, November, 1915.

CHAPTER IX

DECLINING YEARS

Lunar Society—Illness—Death—Funeral—Character—Appearance—Portraits—Distinctions—Will—Epilogue.

IN a previous chapter we mentioned how Boulton's hospitality and the pleasure he took in intercourse with educated men, especially of a scientific turn of mind, led to the formation in Birmingham of an association of such men who met at one another's houses and discussed all manner of subjects in the heavens and on the earth. The nucleus was undoubtedly Boulton, Darwin and Small. As time went on and their numbers increased, a more or less regular time of meeting was chosen—i.e. the time of full moon so that the guests might have the benefit of its light when returning home afterwards. From this fact the name Lunar Society came to be given to the association. As far as we can discover, no formal constitution was adopted although at one time Boulton drew up some rules for its conduct. One can believe this readily, but one can also believe that his proposals would meet with gentle raillery. It does not appear that any definite work was done by the Society such as we see to-day when we cannot meet without some person being constrained to give a paper. The members were quite content to discuss the latest discoveries in science, new theories and learned publications at home and abroad. But it is not to be supposed that these discussions were confined to serious subjects. A great deal of good-humoured chaff and leg-pulling went on. Darwin must have been the life and soul of the meetings. On one occasion he introduced to the notice of members a speaking machine which he had contrived; it could say "pa pa" and "ma ma" and he evidently boasted that he would not have much difficulty in making the machine speak whole sentences. Boulton good humouredly challenged him to do so and drew up the following agreement:[1]

[1] Bolton, H. C., *Scientific Correspondence of Dr Joseph Priestley*, 1892, p. 210; Appendix, p. 195, "The Lunar Society". This is quite the best published account of the Society.

I promise to pay to Dr Darwin of Lichfield one thousand pounds upon his delivering to me (within two years from date hereof) an instrument called an organ that is capable of pronouncing the Lord's Prayer, the Creed and the Ten Commandments in the vulgar tongue, and his ceding to me, and me only, the property of said invention with all the advantages thereunto appertaining.

This was signed by Boulton and witnessed by Small and Keir. Darwin to his regret moved in 1782 from Lichfield to Derby and although he still tried to attend was often baulked by his professional duties; hence a characteristic note such as this to Boulton:

I am here cut off from the milk of science which flows in such abundant streams from your learned lunations; which I can assure you is a very great regret to me.

Or this note on another occasion:

Lord! what invention, what wit, what rhetorick, metaphysical, mechanical and pyrotechnical will be on the wing, bandied like a shuttlecock from one to another of your troop of philosophers.

But if the Society lost a clever man in the person of Darwin, it gained another by the arrival in Birmingham in 1780 of Dr Joseph Priestley.

It is not necessary for our purpose to detail his previous career; we need only say that he had discovered oxygen in 1774 and that he was the best known experimenter of the day in the domain of pneumatic chemistry. He was at once drawn into the circle. He inspired the members with a desire to pursue his own line of investigation and the study of chemistry became all the rage. Boulton was attacked with the fever worse than any one else as witness this letter (1781, Sept. 6) to Lieut. L. Henderson, one of the assistants of Boulton and Watt:

Chemistry has been for some time my hobby horse but I am prevented from riding it by cursed business except now and then of a Sunday. However, I have made great progress since I saw you and am almost an adept in metallurgical moist chemistry. I have got all that part of Bergmann's last volume translated and have learnt from it many new facts.

Boulton goes on to say that he is fitting up a regular chemical laboratory. It will be recalled that when he was in Cornwall in 1779, he fitted up one there.

Several other members like Wedgwood and Darwin shared Boulton's enthusiasm and we can say that if they advanced the cause of science in no other way, they did so, to use a modern phrase, by endowing research. They took the practical step of raising among themselves a subscription to free Priestley from money worries so that he might devote himself to his experimental work. Wedgwood wrote (1781, Mar. 10):

> It would be a pity that Dr Priestley should have any cares or cramps to interrupt him in the fine vein of experiments he is in the midst of, and is willing to devote his time to the pursuit of, for the public good.

It was these experiments, be it recalled, that led on Watt and others to the discovery of the composition of water.

The aim of Wedgwood was to collect £100 per annum for the Doctor, and this they succeeded in doing without hurting his feelings and without making any member feel that he was not giving as much as some one else, by limiting subscriptions to ten guineas. This incident reflects the greatest credit upon the members.

Besides the regular members of the Society, although perhaps that is not the right adjective, numbers of distinguished men were guests on occasion. Indeed, one such could hardly visit the neighbourhood of Birmingham without receiving an invitation to be present. We must mention a few: John Smeaton (1724–92), the engineer and a quondam colleague of Watt; Dr Joseph Black (1728–99), Watt's patron; Sir Joseph Banks (1743–1820), President of the Royal Society; Sir William Herschel (1738–1822), the astronomer; Dr D. C. Solander, who has already been mentioned; and J. A. de Luc (1727–1817), natural philosopher and friend at Court. The proportion of Fellows of the Royal Society among the members and guests of the Society was remarkably high.

The Revolution in France in 1789 caused quite a ferment in this country among all classes. Every thoughtful person was deeply stirred by the event. We have seen already how young Watt was affected. Dr Priestley was no less moved; he forgot for the nonce his scientific studies to preach and write on the brotherhood of man and the downfall of tyranny and priestcraft. By this conduct and by his religious belief he incurred the antagonism of the less thinking part of the community. An incident occurred which was turned into a weapon against him. On the second anniversary of the Revolution, July 14th, 1791, a dinner to celebrate it was held at the principal inn in Birmingham, James Keir being in the Chair. This aroused the resentment of the mob which assembled outside the inn, broke the windows and smashed the furniture. Although Priestley had not been at the dinner, the mob went with shouts of "Church and King" to the unitarian chapel where he laboured. This they gutted and set fire to. Although it was now late at night, the mob set off to Priestley's house at Fairhill. He had received notice of their coming and he with his family escaped in the nick of time. The mob sacked the house and fired the contents including his scientific manuscripts, the result of twenty years' labour. For days afterwards half-burnt papers were fluttering about the spot. Other members of the Lunar Society were naturally suspect, Boulton especially, and it was only the distance from Birmingham of Soho that saved it from attack. Boulton barricaded the doors and armed his trusty workmen with muskets to resist the attack which fortunately never took place. These muskets can still be seen in the Boulton and Watt Collection. It was only after a five days' orgy that order was restored by the arrival of the military. The loss to Priestley was severe. He removed to London temporarily and three years later he emigrated to Pennsylvania, settling in Northumberland County whence he wrote to Watt recalling "the pleasing intercourse I have had with you and all my friends of the Lunar Society. Such another I can never expect to see—indeed London cannot furnish it. I shall always think of you at the usual time of your meetings." Priestley died in 1804 and alas! there were many other lapses from the circle. The advance of old age and the deaths of members—Darwin, Boulton's oldest friend, died in 1802—gradually brought about the dissolution of the Society. As far as we can gather, after Boulton, the choice spirit that had held the Society together, was gone, no more meetings were held.

Boulton had a splendid constitution and with the exception of minor ailments enjoyed good health till nearly the end of his life. By the time he had passed the Psalmist's allotted span, he began to show signs of that infirmity—stone in the kidney—from which he suffered acutely before the end. Nature herself as well as his friends gave him warning that he ought to retire. Such a course was entirely alien to his nature, however. He was one of those men who will wear out but cannot consent to rust out. With mental powers unimpaired, he would have been entirely at a loss had he had no occupation. His hobby was his work—his beloved Manufactory and especially the Mint. When he felt that he needed a change he would go off to Cheltenham to take a course of the waters there, which always benefited him. Even there he could not be idle; on one of his visits he noted an analysis of the waters which he duly entered in his memorandum book.

Medical science of those days was not equal to do more than alleviate the complaint from which Boulton suffered. He spent a great deal of time confined to his room and suffered much pain but without a murmur. In the postscript to the letter (1802, Mar. 29),[1] previously quoted, addressed to Sir Stephen Cottrell of the Board of Trade, Boulton says:

> I have been confined to my Chamber for a fortnight past but am better today & am directing a grand illumination for the peace. I hope I shall not be call'd to Town at present, because my phisicians have prohibited me from travelling further than from my Bed to my Couch.

The occasion was the celebration of the short-lived Peace of Amiens which at Soho was signalized by the illumination of the Manufactory by coal-gas light. This, the invention of William Murdock, was being taken up at the Foundry. The illumination was so novel that it excited the utmost astonishment and admiration. On such an occasion we can be sure that Boulton

[1] Record Office, Board of Trade, 6/118, No. 40.

would not have failed to appear to lead the rejoicings, even if only for an hour or two, and would have defied the orders of any physician not to do so.

The opportunity afforded by the Peace was taken by Mr and Mrs Watt to make an extended tour on the Continent. While away Boulton, hopeful as ever, wrote to Watt:

It is necessary for me to pass a great part of my time in or upon the bed; nevertheless I go down to the Manufactory or the Mint once or twice a day without injuring myself as heretofore but not without some fatigue. However, as I am now taking bark twice a day, I find a daily increase of strength, and flatter myself with the pleasure of taking a journey to Paris in April or May next.

Mr and Mrs Watt on their return from their Continental trip were met in London by a letter of welcome from Boulton, conveying, however, the sad news of the death of Mrs Keir, the wife of their partner. She was a frequent visitor to Soho and a favourite of Boulton, but she was only one of the many old friends whose passing was reducing little by little the Soho circle.

Boulton's days for travelling were, however, at an end. The farthest he could go was to the Manufactory; indeed it helped to divert his mind from the ever-recurrent pain that he suffered. From the beginning of March 1802 to Christmas Eve of that year he was confined to the house.

In 1807, there was great controversy in Birmingham over the question of whether to have a theatre or not. There was a strong nonconformist body of opinion that was greatly opposed to the theatre on the ground that it was immoral. The liberal-minded Boulton was, as one would expect, strongly in its favour and his influence carried the day, with the result that in October a "Docket of licence under his Majesty's sign manual subscribed by the Attorney General for Matthew Boulton and others to establish a theatre in Birmingham"[1] was received. The debt of the inhabitants to Boulton for his public-spirited efforts in this respect is too little known.

Towards the end of 1807 Boulton had another serious attack of his disease; he rallied once more but with less resilience than before. In March, 1809, he was again struck down and every-

[1] MS. in Birmingham Reference Library.

one, including himself, thought the end had come, as the following letter written to his daughter (1809, Mar. 11) shows. The handwriting is so shaky that it is with difficulty decipherable. We reproduce it because of its melancholy interest for it was the last he ever wrote.

It is with difficulty that I write this which you cannot be assured that I shall not be hapy untill I clasp you in my arms for I am now very very mis[er]able & therefore beg you will return to Birmingham where I hope to receive you in 7 or 8 days if you wish to see me living. Pray come soon for I am very ill.

Nevertheless he lingered on for five months more, surrounded with every care and attention, and expired peacefully on Thursday, August 17th, 1809, in the eighty-first year of his age. His death, though of course fully expected by his family, came as a shock to the large circle of his friends, his business associates, and his workmen by whom he was admired and, we can truthfully say, loved.

His son Matthew Robinson decided that the funeral should be "furnished in the handsomest manner avoiding ostentation" and entrusted the arrangements to Mr George Lander of Birmingham. He interpreted the instructions liberally and the lavish manner in which the funeral was carried out may be judged by the fact that his bill amounted to £544. 17*s*. 2*d*.

The funeral was fixed for the following Thursday, August 24th, and was attended by a large concourse of mourners from far and near. The hearse and nine mourning coaches were followed by a procession of 600 workmen from the Manufactory and the Foundry while the line of route of the procession from Soho House to the Parish Church of St Mary, Handsworth, was lined by thousands of spectators. The coffin was borne into the Church on the shoulders of the oldest of the workmen. The service, conducted by the rector of the parish, the Rev. T. L. Freer, was a choral one, or as described on a printed order paper (Copy in the B. and W. Coll.) a "Solemn Musical Service", the hymns and other selections of sacred music being sung by a choir of the Birmingham Society of which Boulton was a patron. It seemed as if everyone sought to do honour to the memory of the great man. To each individual who was invited to take part in the funeral, a memorial medal in copper, the work of John Phillp, was presented. On the obverse the medal bore the words: "Matthew Boulton, died August 17th 1809 Aged 81 years" and on the reverse, in a wreath of palm, the words: "In memory of his obsequies August 24th 1809". It was no small feat to have got this done within a week.

It is a matter for regret that Mr Lander's conduct of the funeral came under the suspicion of young Boulton who declined to pay some £50 odd of the amount of the bill. In fact this led to a lawsuit, Lander *v.* Boulton, and some undignified pamphleteering, but eventually the bill was paid in full.

The obituary notices of Boulton were many and were just in their appreciation of his work; we need quote only one:[1]

> His life has been an uninterrupted application to the advancement of the useful arts, and to the promotion of the commercial interests of his native country.... Such was the man who constituted the mainspring to various and extensive ramifications of the mechanical arts unknown to former times.

Boulton's son placed a bust of his father, executed by John Flaxman the sculptor, in the chancel of the Parish Church, accompanied by a mural tablet, the inscription on which is notable for its verity and for the dignity of the language in which it is expressed, in a day when lapidary inscription was indeed raised to a fine art. The wording is as follows:

Sacred to the Memory of

MATTHEW BOULTON, F.R.S.

By the skilful exertion of a mind turned to Philosophy and Mechanics
The application of a Taste correct and refined
And an ardent Spirit of Enterprize, he improved, embellished and extended
The Arts and Manufactures of his Country,
Leaving his establishment of Soho a noble Monument of his
Genius, Industry and Success.
The Character his talents had raised, his Virtues adorned and exalted.
Active to discern Merit and prompt to relieve Distress
His Encouragement was liberal, his Benevolence unwearied
Honoured and admired at home and abroad
He closed a life eminently useful, the 17th of August 1809 aged 81
Esteemed, loved and lamented

[1] *Birmingham Herald*, August 18th, 1809, and copied by other papers.

Watt had a hand in this inscription and it does credit to his head and heart. He wrote too a letter of condolence to young Boulton from Glenarback in Scotland where he was staying (1809, Aug. 23). We quote a passage:

> We may lament our own loss, but we must consider on the other side his relief from the torturing pain he has so long endured & console ourselves with the remembrance of his virtues & eminent qualifications. Few men have had his abilities & still fewer have exerted them as he has done and if to them we add his urbanity, his generosity and his affection to his friends, we shall make up a character rarely to be equalled. Such was the Friend we have lost and of whose affection we have reason to be proud, as you have to be the son of such a Father!

A month later at the son's request Watt penned the *Memoir* from which we have quoted so freely. Its importance is such that we have deemed it right to print it as an Appendix.

In his private relationships, Boulton was a man of deep affections demanding in return love and sympathy; as we have said he was fond of the society of women and they were equally fond of his, while his men friends felt that he was almost more than a brother to them. Even the undemonstrative Watt could say: "he was a most affectionate & steady friend & patron with whom during a close connection of 35 years I have never had any serious difference." Boulton was fond of children and was a prime favourite among the young folks. We have a fine instance of this in his consideration for James Watt, junior, who when he was in Manchester in 1789 on the threshold of his business career got into debt, because the allowance his father made him was insufficient to permit of his associating with the company to which his father's reputation had given him an introduction. The young man turned for help, not to his less comprehending father, but to the more sympathetic Boulton with the happy result that on Christmas Eve arrived the welcome present of a draft for £50 accompanied by a letter of fatherly advice from Boulton.

By the workmen and workwomen in his employ he was looked up to and loved for he lived among them and was approachable in the old-fashioned way; he was, as Boswell so well describes him, like the chieftain of his tribe.

Of Boulton's hospitality we have already remarked upon the profuseness. In the case of visitors to the Manufactory, this might be said to have had an ulterior motive, but in the case of his friends it was an expression of his inmost feelings. Indeed, it can hardly be denied that it was to this trait in his character that the Lunar Society owed its being.

He was fond of the good things of life but we do not hear of any over-indulgence in food or wine as was so common in his day. To excesses in others he was tolerant; while Watt would have turned off a man for laziness or drunkenness Boulton could always advance some excuse for him.

Of his benevolence to those in distress we have given instances such as the cases of the Fothergill family and of the widow Swellengrebel, and these must suffice.

Boulton had his faults—who has not?—but we shall be to them a little kind and say nothing of them.

Looked at from the business side, we should say that Boulton was what we term in modern parlance an *entrepreneur*. His own words to Watt in 1769, when negotiations were in progress to take up the latter's steam engine, express Boulton's motive in whatever he undertook: "love of a money-getting ingenious project". The order of the adjectives should, however, be reversed: the project had to be first and foremost "ingenious" to enable him to exercise sufficiently his eminently agile and inventive brain; "money-getting" was only important to him in that it afforded the wherewithal to launch out into further schemes. His own inventiveness and ingenuity have been referred to already.

Watt's estimate of his partner's abilities, in the *Memoir* so frequently quoted, is so acute and so judicial that the author cannot better it by any words of his own. It is as follows:

Mr Boulton was not only an ingenious mechanick, well-skilled in all the practices of the Birmingham manufacturers, but possessed in a high degree the faculty of rendering any new invention of his own or others useful to the publick, by organizing & arranging the processes by which it could be carried on, as well as of promoting the sale by his own exertions & by his numerous friends & correspondents. His conception of the nature of any invention was quick & he was not less quick in perceiving the uses to which it might be applied, & the profits which might accrue from it. When he took any scheme in hand, he was rapid in executing it, & on these occasions neither spared his own trouble [n]or expence. He was a liberal encourager of merit in others, & to him the country is indebted for various improvements brought forward under his auspices which have not been mentioned.

From what has now been said the reader will have formed already an estimate of Boulton's character. His temperament was extremely sanguine, he was impulsive, buoyant, hopeful and greatly daring, but this did not cause him to be reckless for he had that slight admixture of the phlegmatic which made him look before he leapt. His abilities were of a high order. He had a power of organization rarely met with, love of work, promptitude, perseverance, and intense application. He was gifted with extraordinary tact in dealing with his fellow-men—he could always "find a way", perhaps a compromise, for he was intensely practical; what assisted him greatly in this respect was an uncanny power of reading character quickly. Thus he could "size up" a man or pick out the right one for the job.

His word was his bond and everybody seems to have trusted him implicitly. As testimony on this point, we would point to the fact that, extraordinary as it may seem, legally executed deeds of his various partnerships with Fothergill, Watt and others are not to be found. There may be some deep reason for this, such as the law of partnership of that day whereby a partner's liability was unlimited so that a man might be ruined by a single venture that turned out unsuccessful.

Boulton was never satisfied to do anything short of the best. To quote Keir again: "To understand perfectly the character of Mr B's mind, it is necessary to recollect that, whatever he did or attempted, his success or his failure were all on a large scale."

Boulton mentally was very acute and "quick in the uptake".

He gleaned a great deal of his knowledge by intercourse with the world and with his friends. As Keir said of him in the memoir prepared in 1809:

Mr B. is a proof of how much scientific knowledge may be acquired without much regular study, by means of a quick & just apprehension, much practical application, and nice mechanical feelings. He had very correct notions of the several branches of natural philosophy and was master of every metallic art, & possessed all the chemistry that had any relation to the objects of his various manufactures. Electricity and astronomy were at one time among his favorite amusements. It cannot be doubted that he was indebted for much of his knowledge to the best preceptor—the conversation of eminent men.

To sum up, Boulton was a man of truly noble character, high principled and generous. Watt expressed it well when he spoke of him as the "princely Boulton". He was the highest type of Englishman—one of those who sustain in their day and generation the credit of this nation for integrity and fair dealing.

In appearance Boulton was above the medium height with a fine figure and erect carriage. He had a handsome face, with somewhat receding forehead, a firm chin and grey eyes with a humorous twinkle in them under well-arched eyebrows. One contemporary writer, a disciple of Gall, states that "his organs of comparison, constructiveness and of individuality were immense" but this was a *post hoc* observation.

Boulton's manners were charming and easy as of one accustomed to wealth and habitual command. Doubtless this arose from his intercourse with persons in high places in a period when manners certainly were polished. This address secured for him the entrée to the very highest in the land. For this we have the testimony of Keir, in his memoir, in these words:

Perhaps no man in his station had ever been received by the great so graciously & with less of that etiquette which constrains the freedom & ease of conversation. Even the highest person in the kingdom received him with much condecention & gave him access to the palace whenever he, Mr B., wished to shew any new piece of workmanship which he thought might give pleasure to his Sovereign. It is even said, not without authority, that when his Majesty intended an excursion into Warwickshire & some neighbouring counties, he mentioned his intention of seeing Mr B. & being told that Mr B's severe & dangerous illness would prevent him from receiving the high honour intended, his Majesty replied with a sentence expressive of his goodness & greatness of mind "that he should visit Mr B in his sick chamber". If we agree with Horace, *Principibus placuisse viris non ultima laus est*, few men have been entitled to this commendation.

Boulton was always well dressed and we must not forget what an addition to a man's appearance was afforded by the picturesque dress of the period: the grey peruke, the embroidered coat, set off with some of his own buttons, the lace jabot and lace at the wrists, the flowered waistcoat, the knee-breeches, the silk stockings, the inlaid buckles, again his own, on the polished shoes; we hear even, on one occasion, of his wearing a sword.

A number of contemporary portraits exist and help us to visualize his appearance. The earliest and the one selected for reproduction in this volume is the oil painting by Charles Frederick von Breda, R.A., 1792, now in the possession of the Institution of Civil Engineers. The most virile portrait, and in the opinion of many the best, is that by Lemuel Abbott in the possession of the City of Birmingham; it is not known when this was painted. The best known portrait is that painted by Sir William Beechey, R.A., in 1801, in the possession of the Boulton family. He was then past his prime and the artist's representation of him as a benevolent old gentleman, while true at the time it was painted, does not convey the character of the man of action whom these pages have revealed.

Several good busts of Boulton exist. We have mentioned the one by Flaxman in the chancel of Handsworth Parish Church. The busts shown on the Boulton memorial and other medals,[1] one of which is reproduced on Plate X, are excellent. Several striking portraits in flesh-coloured wax by Peter Rouw, mostly of date 1803, exist. Posthumous portraits are based on the foregoing material.

[1] Cf. pp. 162, 202, and Eidlitz, R. J., *Medallic Portraits of Matthew Boulton and James Watt*, Privately printed, 1928. 4to.

Many distinctions fell to Boulton's lot. He was High Sheriff of Staffordshire in 1795. The Letters Patent dated February 5th, 1794, "committing to him the custody of the county of Stafford", are still preserved.

We may perhaps call membership of the Lunar Society a distinction; at any rate it was the forcing ground that led to his election with Watt in 1783 as a Fellow of the Royal Society of Edinburgh. Their membership badges are preserved in the Boulton and Watt Collection. Fellowship of the Royal Society, London, along with Watt, followed in 1784; their signatures appear together on the Roll of Fellows. Surprisingly enough we do not find, as we should have expected, that Boulton was a Member of the Society for the Encouragement of Arts, Manufactures and Commerce, to give it its full title. In this respect he was in the same boat as Wedgwood. Did they consider it dilettante?

In 1792, Boulton became an Honorary Member of the Society of Civil Engineers founded by John Smeaton in 1771, the first of its kind in the world and the parent of subsequent engineering societies. It is a satisfaction to know that it is still in existence under the title "Smeatonian Society of Civil Engineers".

Boulton was one of the original proprietors of the Royal Institution of Great Britain; he was elected in 1800, that is shortly after its foundation by Count Rumford. Until the Act of Parliament amending its constitution was passed in 1810, there were only Proprietors, and Subscribers to the lectures, so that Boulton was not a Member in the sense that we understand it to-day. It does not appear that he took any active part in its affairs, but this can be explained by the state of his health which in his latter years prevented journeys to town.

It is stated that Boulton was a member of the Voluntary Economical Society of St Petersburg. This, the oldest scientific Society in Russia, for it was founded in 1765, somewhat resembled in scope our Society of Arts mentioned above (founded 1754) except that it was more specifically interested in agriculture and associated pursuits.[1] It survived till 1917.

[1] For this information the author is indebted to Prof. P. P. Zabarinsky of the Institute for the History of Science and Technology of Leningrad.

Boulton was armigerous or perhaps it would be safer to say that he bore arms. His coat was: Azure, on a bend or, cottised argent, between two fleur-de-lys of the second, an anchor sable between two leopards' faces of the field. His motto was "Faire mon devoir" and we say unhesitatingly that he lived up to it. Boulton had a seal with these arms, the earliest use of which the author has traced was in 1778. He is informed by the Lancaster Herald that Boulton did not take out a grant of arms at the Heralds' College. Presumably therefore he bore these arms by inheritance. To those skilled in such matters, this would be a clue to his descent, or what he believed to be his descent, farther back than we have been able to establish (see the pedigree, p. 27).

Boulton's son, as a result of his father's marriage with an heiress, bore these arms quartered with those of Robinson—vert on a chevron or, between three bucks trippant erminous, as many trefoils of the field—and these are the arms borne by the family to-day.

A man's last will and testament is frequently a self-revealing document, and so it was in the case of Boulton, for it shows that he had given much thought to remembering relations, friends and servants. His will[1] is dated June 23rd, 1806, and by the way one of the witnesses is James Watt. Boulton left his estate, subject to various bequests in a schedule attached, to his son Matthew Robinson whom he appointed sole executor. For his unmarried daughter Anne he made a settlement of £5000 and appointed James Watt, junior, James Simcox and Nathaniel Gooday Clark executors under the settlement, leaving each £50 for his trouble. He bequeathed to his nephew Zacchaeus Walker £500, and to his niece Ann Walker £50; to his nephew George Nunn £500 and to his niece Mary Mynd £200. (To what branch of the family the latter two belonged we have not been able to find.) Among the general bequests the more interesting are those of £100 each to John Thomas and Charles "sons of my late partner John Fothergill", Capt. William Fothergill, R.N., another son probably, received a like

[1] Somerset House, P.P.C. 1807, 672 Loveday.

sum. Miss Frances de Luc, daughter of his friend J. A. de Luc, was remembered to the amount of £50. Miss Amelia Alston, whom we believe to have been housekeeper at Soho House, received £100, "Mrs Mary Keen, the elder of Stafford", who may perhaps have held the post previously, a like sum. Three clerks in the businesses benefited: Andrew Collins to the amount of £20; William Dodding £200, and James Pierson £50. An intimate touch is given by the bequest of £110 to "William Sandbrook my domestic"—no doubt his valet during his illness—and by the bequest of £20 to "Joseph Brooks my coachman". The £20 left to "John Bush one of my workmen" was well deserved we can be sure; he had been employed as we have seen on the coining machinery. Nor were the managers in the works forgotten: John Southern received 30 guineas for the purchase, very appropriately, of philosophical instruments; James Lawson and William Murdock were each left a like sum. Ambrose Weston, Boulton's legal adviser in London, and James Weston his brother were left £100 each. John Woodward and John Mosley—probably clerks in London —had £50 each. Quite in the fashion of the day, but none the less a kindly thought, was the bequest to John Hodges of a sum of £20 for a suit of mourning and 10 guineas to Croat Dixon for a like purpose.

There is a codicil to the will dated October 7th, 1807, whereby Boulton leaves on trust £100 to the General Hospital, Birmingham, and a further like sum for the Dispensary; of this bequest he appointed Heneage Legge and Samuel Galton trustees. By a further codicil dated March 18th, 1808, he left to his son as heirlooms the diamond ring presented to him by the Emperor Alexander of Russia and a ring presented on the death of Lord Nelson.

The will with codicils was proved in London by the executor, September 15th, 1809, for £150,000, a very large sum for those days but not disproportionate to the magnitude of Boulton's enterprises; truly he had not buried his talent in a napkin.

Matthew Boulton has not been forgotten entirely, although

his birthplace still lacks a monument to him. We trust the public spirit of its citizens will see to it that this defect shall not go much longer unremedied. The centenary of his death was marked, we are pleased to say, by a memorial service in the Parish Church of Handsworth on September 15th, 1909, when the sermon was preached by the Rev. Prebendary Burn, D.D. Five of Boulton's great-grandchildren were present.

We have now exhibited Boulton as a great captain of industry in every successive phase of his life: toy-maker, artist, engineer and coiner. In all these employments he excelled, but his great legacy to the world, a legacy we are enjoying to-day, was the steam engine of which he was, as he foreshadowed he would be, nurse and midwife.

APPENDIX I

MEMOIR OF BOULTON BY WATT

Holograph

Glasgow, September 17th, 1809. Memorandum concerning Mr Boulton, commencing with my first acquaintance with him.

I was introduced at Soho by Dr Small in 1767 but Mr Boulton was then absent; Mr Fothergill his partner & Dr Small showed me the works. The goods then manufactured there were steel gilt & fancy buttons, steel watch chains & sword hilts, plated wares, ornamental works in Or moulu, Tortoise shell snuff boxes, Bath metal buttons inlaid with steel & various other articles which I have now forgot. A mill with a water wheel was employed in Laminating metal for the buttons, plated goods &c and to turn Laps for grinding & polishing steel work & I was informed that Mr Boulton was the first Inventor of the inlaid buttons, & the first who had applied a mill to turn the Laps. Mr B. at that [time] also carried on a very considerable trade in the manufacture of buckle chapes, in the making of which he had made several very ingenious improvements. Besides the Laps in the mill, I saw an ingenious lap turned by a handwheel for cutting & polishing the steel studs for ornamenting buttons, chains, sword hilts &c, and a shaking box put in motion by the mill for scowering button blanks & other small pieces of metal which was also a thought of Mr B.—there was also a steelhouse for converting iron into steel, which was frequently employed to convert the cuttings & scraps of the chapes & other small iron wares into steel which was afterwards melted & made into cast steel for various uses.

In 1768 I was again at Soho, on my return from London where I had been taking the necessary steps for obtaining a patent for the improved steam engine, & was then introduced to Mr Boulton, by Mr Garbett or Dr Small. I found the same manufactures carried on & had much conversation with Mr B. on them when he explained to me many things of which I had been before ignorant. On my part I explained to him my invention of the Steam Engine & several other schemes of which my head was then full, in the success of which he expressed a friendly interest. My stay at Birmingham at that time was short, but I afterwards kept up a correspondence with Mr B. through our mutual friend Dr Small.

In 1773 or 1774, finding that in my then situation & circumstances I could not bring my invention into general use, I prevailed upon Dr Roebuck who was then my associate & Patron to offer his share in the invention to Mr B. on certain terms, which after some hesitation, he accepted, provided he was satisfied with the invention on trial at Soho. I accordingly sent to him an Engine with an 18 inch cylinder, which I had erected some years before at Kinneil in Scotland & followed it myself in the summer of that year. After sufficient trial Mr B. agreed to assist & support me in obtaining an act of parliament for the prolongation of the patent, which by his exertions, the assistance of Dr R & all the friends we could muster was obtained after a long attendance upon Parliament; after which Dr R. was induced to transfer all his rights to the profits to Mr B. in consideration of certain sums paid & to be paid him & Mr B. entered into partnership with me, his partner Mr Fothergill having some time before declined taking any share. Through the whole of this business Mr B's. active & sanguine disposition, served to counterballance the despondency & diffidence which were natural to me & every assistance which Soho or Birmingham could afford was procured. Mr B's amiable & friendly character together with his fame as an ingenious & active manufacturer procured me many & very active friends in both houses of parliament.

The first large engine we made was at Bloomfield Colliery in Staffordshire & nearly at the same time one for Mr Wilkinson at Broseley Iron works, to blow bellows; soon after we erected one at Hawkesberry Colliery & a small one at Messrs. Cooke & Comps. Distillery at Stratford Le bow, which was followed by others in various places & among the rest by engines at Shadwell & Chelsea waterworks.

I think it was in 1778 we erected the first engine in Cornwall on an addit belonging to Chacewater mine, which was soon followed by Tingtang & Poldice engines. The order & even names of all the engines we erected in that county I cannot now recollect, but it was in 1781 that we erected 5 engines on Wheal Virgin & the Consolidated Mines. The series of our erections in Cornwall & other places it is unnecessary now to enumerate.

I think it was in 1782 that I obtained a patent for the rotative engines which with their subsequent improvements have been of so much service to the manufactures of this country. The first in London was erected for Messrs. Goodwyn & Co., Brewers, which was followed by Mr Whitbread & others. But what raised their reputation the most was their splendid exhibition at the Albion mill, a concern highly beneficial to the publick & by its catastrophe highly injurious to its owners. I forbear tracing the steam engine farther lest I should appear to be writing its history instead of that of Mr Boulton. Suffice it to say that to his generous patronage, the active part he took in the management of the business, to his judicious advice & to his assistance in contriving & arranging many of the applications to various machines, the publick is indebted for great part of the benefits they now derive from that machine; without him, or some similar partner (could such a one have been found) the invention could never have been carried by me to the length it has been.

C

Soon after my connection with Mr B, he declined the or moulu business & the tortoise shell boxes, but supported Mr Francis Egginton in the manufacture of what are now called polygraphick pictures, which some time afterwards he resigned entirely to Mr Egginton.

About this time a quantity of the wheel work for the reels used in organizing silk were wanted by the E. India Compy, which Mr Boulton undertook & by the assisstance of the late Mr Rehe made considerable improvements in. That company also wanted a large quantity of a peculiar sort of Tobacco boxes which Mr Boulton contracted for at a very low price which he was enabled to do by making them of Bath metal which admitted of being struck when hot in very handsome forms; they could not have been made of brass at twice the money.

B

When the new coinage of gold took place in 17[85] Mr B. was employed to receive & exchange the old coin, which served to revive his ideas on the subject of coinage which he had considered as capable of great improvement. Among other things, he conceived that the coin should all be struck in collars, to make it round & all of one size which is by no means the case with the common gold coin & if also all of one thickness, the purity of the gold might be determined by passing it through a guage or slitt in a piece of steel made exactly to fit it; and he accordingly made a proof guinea, with a raised border & the letters en creux, somewhat similar to the penny pieces he afterwards coined for Government, & which completely answered the intention, as any piece of baser metal which fitted the guage was found to be considerably lighter & if made to the proper weight would not go through the guage. Such money was also less liable to wear in the pocket than the common coin where all the impression is prominent. His proposals on this head were not however approved of by those who had the management of his Majesty's mint, & there the affair rested for the time.

In 1786 Mr B. & I were in France where we saw a very fine crown piece executed by Mr P. Droz in a new manner. It was coined in a collar split into 6 parts which came together when the dies came into contact with the blank & formed the edge & the inscription upon it. Mr D. had also made some other improvements on the coining press & pretended to others in the art of multiplying the dies. And, as to his mechanical abilities Mr D. joined that of being a good die sinker, Mr B. contracted with him to come over at a high salary & work at Soho, Mr B. having then a prospect of an extensive copper coinage for the East India Comp[y] & a probability of one for Government. A number of coining presses were constructed & a Steam Engine applied to work them. Mr Droz was found to be of a troublesome disposition, several of his contrivences were found not to answer & were obliged to be better contrived or totaly changed by Mr B. & his assistants. The split collar was found to be difficult of execution, & subject to wear very soon when in use, & in short very unfit for an extensive coinage. Other methods were therefore adopted & it was laid to rest. Mr Droz was dismissed after being liberally paid, and other engravers were procured (It should be noticed here that Mr D's method of multiplying the Dies did not answer & that it appeared that he did not know so much on that subject as Mr B. himself did. That process was therefore brought to perfection by Mr B. himself & his assistants)—Much ingenuity, time & great expence were required to perfect the application of the steam engine to coining, in all which Mr B. acted the principal part & gave life to the whole. The machinery being perfected, Mr B. undertook & executed several extensive coinages both for Government, for the [East] India Company & others of minor importance & to his exertions are owing the perfection the copper coin of this country has now attained to. He also executed a considerable quantity of beautiful coin for the revolutionary government of France, while we remained at peace with that country, which coin was afterwards suppressed by the arbitrary measures of a fresh set of rulers in that unhappy country, to the great Loss of the French contractors who nevertheless paid Mr B. honourably.

Although the expensive machinery which has been mentioned performed very well, yet Mr B. was not satisfied, and some years afterwards he set about & executed machinery for that purpose, in a manner which does not seem to admit of any material improvement, & with which he has executed several large coinages.

In the year [1799] Mr B. with the approbation of our government contracted with the Emperor (Paul) of Russia for a complete coining apparatus, which is now at work at St Petersburgh, and after that with the Danish government for one which is erected at Copenhagen. Since that time he has been employed by our government to erect one on Tower hill, which is nearly completed & will be one of the finest establishments of the kind which have ever been executed & to the planning & establishment of which he attended even under the pressure of age & of a painful disease. In short had Mr B. done nothing more in the world than what he has done in improving the coinage, his fame would have deserved to be immortalized, & if it is considered that this was done in the midst of various other important avocations, & at an enormous expense for which he could have no certainty of an adequate return, we shall be at a loss whether to admire most his ingenuity, his perseverance or his munificence. He has conducted the whole more like a sovereign than a private manufacturer. The Love of fame has been to him a greater stimulus than the love of gain; yet it is to be hoped that even in the latter view, it has answered the purpose.

To enter into all the various mechanical improvements which have been due either to his own invention or to his fatherly patronage is beyond my powers; of some I may never have heard, & others may have slipt my memory.

I have omitted to mention (A) that the first engine of 18 inch cylinder, which was employed in returning the water to Soho Mill was replaced about 1778 or 79 by a larger engine, the first on the expansive principle, which still remains there. And that (B) there were two different orders for the E. India reels, in the first of which I think Mr John Whitehurst of Derby made the model & gave other assistance, and in the second order Mr Rehe directed the execution. At this time also our very respectable friend Mr Keir being disengaged from other Business, and Mr B. being obliged to be frequently absent, Mr K. gave his assistance in the general superintendance of all the businesses at Soho during which he made many valuable arrangements, & gave other assistance.

In 17[88] Mr B. erected a new rolling or laminating mill with a powerful water wheel, and which also turned Laps & polishers of various kinds & other machinery, in which it is assisted by a rotative steam engine. This mill and engine also turn machinery for cutting out the blanks for coining & performing other operations on them, by very ingenious contrivances, of Mr Boulton's. In the method of forming the rollers or cylinders which laminate the metal & in other points conected with it much merit is due to Mr B.

I have omitted to mention Soho Foundery, as the part which

Mr Boulton took in that is better known to others than to me.

I should have mentioned on the head of the Albion Mill (C) that it was planned by Mr Boulton & the late Mr Saml Wyatt a very ingenious architect & that at that time no edifice of the kind had been constructed with similar conveniences or powers. The machinery was constructed under the direction of Mr Rennie whose abilities as an Engineer are now well known to the publick.

Mr Boulton was not only an ingenious mechanick, well skilled in all the practices of the Birmingham manufacturers but possessed in a high degree the faculty of rendering any new invention of his own or others useful to the publick by organizing & arranging the processes by which it could be carried on, as well as of promoting the sale by his own exertions & by his numerous friends & correspondents.

His conception of the nature of any invention was quick & he was not less quick in perceiving the uses to which it might be applied & the profits which might accrue from it.

When he took any scheme in hand, he was rapid in executing it & on these occasions neither spared his own trouble [n]or expence. He was a liberal encourager of merit in others & to him the country is indebted for various improvements brought forward under his auspices which have not been mentioned.

To his family he was a most affectionate parent. He was steady in his friendships, hospitable & benevolent to his acquaintances & indeed I may say to all who came within his reach who were worthy of his attention & to sum up humane & charitable to the distressed.

I have said he had many friends; some of the most intimate were the late Dr E. Darwin, the late Dr Wm. Small, the late Saml Garbett Esqr, the late Thomas Day Esqr, James Keir Esqr, Chas. Dumergue Esqr & others, which can easily be recollected by his family & friends.

In respect to myself, I can with great sincerity say that he was a most affectionate & steady friend & patron with whom during a close connection of 35 years I have never had any serious difference.

In respect to his improvements & erections at Soho, his turning a barren heath into a delightful garden & the population & riches he has introduced into the parish of Handsworth, I must leave them to those whose pens are better adapted to the purpose & whose ideas are less benumbed with age than mine now are.

J. WATT.

APPENDIX II

BOULTON'S BUSINESSES

The information is compiled from various sources, including Bisset's *Survey of Birmingham*, 1800; the dates are not altogether definite.

Matthew Boulton, 1759–1809.	Mercantile trade in Birmingham.
Boulton & Fothergill, 1762–1781.	Toys, ormolu, plated ware.
Matthew Boulton & Plate Co., ? 1765–1809.	Plated & silver ware.
Boulton & Watt, 1775–1800.	Steam engines.
Boulton & Eginton, 1778–1780.	Copying pictures.
James Watt & Co., 1780–1794.	Letter-copying machines.
Matthew Boulton & Button Co., 1782–1809.	Buttons.
Boulton & Scale, 1782–1796.	Buttons, Buckles.
Matthew Boulton, 1788–1809.	Medals, rolled metals.
Matthew Boulton, 1795–1809.	Mint for government copper coin.
Boulton, Watt & Sons, 1795–1800.	Iron foundry and steam engines.
Boulton & Smith, ?1796–? 1809.	Buckles, latchets.
Boulton, Watt & Co., 1800–1809.	Iron foundry and steam engines.

INDEX

Adam Brothers, influence of, *59*, 185
Adam, James, correspondence with, *60–2*
Adventurers, Cornish, *94–7*, 120, 121
Advertisement, Boulton's belief in, *72*, *89*
Æolus frigate sheathed, 161
Albion Mill, 122–5, 208
Alston, Amelia, 174, 201
American colonies, coinage for, 138
Amiens, Peace of, 190
Anchor mark assigned to Birmingham plate, *69*
Anglesea, copper mines in, 131, 139, 157
Apprentices, Boulton's, *62*, *63*
Argand's lamp, 127–9
Arkwright, Sir Richard, 128, 129
Arms of Boulton, 200
Arrêt de conseil, 98
Art objects collected by Boulton, *55*
Assay Office, Birmingham establishment of, *64–70*
Atmospheric engine, 77, 89
Avery, W. and T. Ltd., 170

Bank of England, coinage for, 155
Banks, Sir Joseph, 188
Barber, Richard, counterfeiter, 152, 154
Barney, Joseph, artist, 105
Baskerville, John, *36*, 88
Beechey, Sir Wm., portraits by, 39, 198
Bentley and Co., engine for, 90
Bersham Iron Works, 85, 158, 164, *166*, *168*
Bewdley, trade at, *16*
Birmingham, early descriptions of, 9, 10; situation of, 19, 20; rise of, 21; enterprise of inhabitants, 21, 22; coming of the Boultons, 24; businesses in, 28, 29; industries in, 31, 32, 41, 42, 48; division of labour in, 41; drawing schools in, 63; illegal coining in, 135, 152–4; theatre in, 195
Birmingham Assay Office established, *64–70*
Birmingham Hospital, legacy to, 201
Birmingham Metal Co., 109
Birmingham and Wolverhampton Canal, 167
Bisset, James, 182, 183, 209
Black, Dr Joseph, 79, 119, 188
Black Country, 8–10
Bloomfield engine, 90, 93, 204
Blowing engine, 90, 204
"Blue john" spar, 57
Boiler, Boulton's steam, 117
Booth, Joseph, polygraphs of, 106–7
Boring mill, Wilkinson's, 85, 88, 92; Soho, 168
Boswell, James, visits Soho, 73
Boulsover, Thomas, invents Sheffield plate, 51
Boulton, Anne, second wife of Boulton, 27, 34, 35, 119
Boulton, Anne, daughter of Boulton, *56*, *166*
Boulton, Mary, first wife of Boulton, 27, 30, 34
Boulton, Matthew, ancestry, 23, 27; birth, 25; schooling, *26*; enters paternal business, 28; marriage, 30; deaths of his wife and of his father, 30; buckle-chape bill, 31–3; second marriage, 34; comes into fortune, *36*; friends and associates, 37–9, 88, 188; buys lease of Soho, 41, 42; establishes Manufactory, 42–5; partnership with Fothergill, 45; makes steel jewellery, 47; makes buttons, 48; enticed to go to Sweden, 49, 50; makes Sheffield plate, 52; makes silver plate, 53, 63–71; makes ormolu, 54; collects art objects, *55*; sale of Soho products, *56*; audience of the King, *56*, 57, 149; manufactures clocks, 57–9; meditates showroom in London, *60–2*; takes apprentices, 62, 63; hall-marking of silver, 64; Assay Offices, *64–70*; founds Birmingham office, *69*; famed for

Boulton, Matthew (*cont.*)
art products, 71; entertains visitors at Soho, 72–4; believes in advertisement, 72, 89; need of further power, 75; attempts a steam engine, 75–6; learns of Watt's engine, 76–7; meets Watt, 79; declines limited licence to make engines, 81; his ideas of manufacture, 81; affected by trade depression, 82–3; takes over Roebuck's share in Watt's patent, 84; applies for extension of patent, 85; commencement of partnership with Watt, 86; his hospitality, 87, 88, 186; brings out the new engine, 89–91; his trust in Watt, 93; introduces engine to Cornwall, 94; in need of fresh capital, 97; engages William Murdock, 97; supplies engines for France, 98–9; meditates engagement of Keir, 100, 101; makes copper alloy for ships' fastenings, 102, 103; copying pictures mechanically, 104–6; takes up Watt's copying press, 108; Birmingham Metal Company, 109; brass manufacture, 109; dissolves partnership with Fothergill, 110; generosity, 111; engages John Scale, 111; clears himself financially, 112; envisages rotative engine, 113; encourages Watt, 115; suggests improvements in boilers, 117; death of his second wife, 118; tours Scotland, 119; Albion Mill, 122; visit to Paris, 124, 136; opinion on patents, 128; loyal address to George III, 130; adventures in Cornish mines, 131; promotes copper trust, 131; suggestions for improving coinage, 133–7; applies steam engine to coining, 138; patents coining press, 140; description of Soho Mint, 142, 157; takes up medallic work, 145–7, 158; executes regal coinage, 148–52, 155; raids coiners' dens, 152–4; makes sheathing for ships, 159–62; education of his son, 164–6; starts Soho Foundry, 167; speech at rearing feast, 169; patents hydraulic ram, 173; establishes Soho Insurance Society, 179–82; apprehends burglars, 183; improves Soho House, 184; studies chemistry, 165, 187; illness, 190–2; supports theatre in Birmingham, 191; death, 192; monument, 193; affection, 194; abilities, 195; character, 196; appearance, 197; distinctions, 199; will 200–1; memoir by Watt, 203
Boulton, Matthew, senior, 24, 25, 27, 29–31
Boulton, Matthew Robinson, infancy, 56; education, 164–6; partner, 167; orders funeral, 192–4; executor, 200
Boulton and Watt, 89–112, 131, 176, 209
Boulton and Watt Collection, xi, 109
Boulton and Watt engines, distribution of, 176–7; prices of, 178
Boulton Papers, xi
Boulton, Watt and Co., 171, 209
Boulton, Watt and Sons, 167, 171, 209
Box, Godfrey, and the slitting mill, 11
Bradley Ironworks, 158
Brass making, 14, 109
Breda's portrait of Boulton, 198
Bridgewater Canal visited by Boulton, 51
Brooks, Joseph, 201
Broseley, engine at, 90
Brummagem products, 51
Buckle making, 15, 16, 31–3, 68, 203
Bull, Edward, litigation with, 173–5
Burglary at Soho, 183
Burke, Edmund, denounces young Watt, 166
Burn, Rev. Prebendary, 202
Bush, John, 147, 148, 201
Businesses of Boulton, 208
Button making, 15, 48, 73, 203

Cadman, George, silver thread edge to Sheffield plate, 52
Calonne, M. de, 124
Camden's account of Birmingham, 10
Cameos, 47

Canal, Birmingham and Wolverhampton, 17, 167; Bridgewater, 8, 50; Staffordshire and Warwickshire, 17; Trent and Mersey, 17
Canal mania, 8
Candelabrum, Sheffield plate, 53; ormolu 57
Candle, self-snuffing, 127
Capper, Peter, 109
Carron Ironworks, 38, 92, 119
Cartel, early instance of, 132
Casements, metal, 103, 185
Cementation steel, 12, 47, 48, 203
Centrifugal governor, 125–6
Chacewater engine, 94, 204
Chamber of Manufacturers, 131
Chape making, 31–3; wages for, 32
Charcoal, 3, 5
Cheadle, brass making at, 14, 109
Chemistry, practical, 165, 187, 188
Christie and Ansell's, sales at, 56, 57
Christmas boxes, 179, 183
Chronicle of Boulton, xiii
Church and King riots, 189
Civil Engineers, Boulton member of Society of, 199
Clark, N. G., 200
Cleanliness in workshops, 143, 181
Clocks made at Soho, 57, 58, 59
Club Rooms at Soho, 181
Coal, fossil or sea, 3, 5, 8, 19
Coal-gas, Murdock's, 190
Coal-tar, Dundonald's, 120
Cochrane, Lord, 119
Coinage, Boulton's ideas on, 135, 148, 205
Coiners' dens, Boulton raids, 152–4
Coining, punishment for illegal, 134, 154; Boulton's improvements in, 138, 140–4, 148, 157, 205
Coining press, 13, 133
Collar, split, for coining, 136, 137, 147, 206
Collins, Andrew, 201
Colquhoun, Patrick, 149, 150
Cook, Adams, Wilbie and Sager, engine for, 93
Cooper, Thomas, 166
Copper mining in Staffordshire, 14
Copying press, Watt's, 107–9, 167
Cornish Metal Company, 132
Corn-milling, steam, 122–4
Cornwall, engines in, 94–7, 99, 113, 120, 121, 204; mining in, 89, 131, 132; riots in, 132; royalties from, 175
Cosgarne, Cornish residence of, 99
Cottrell, Sir Stephen, 190
Counter, engine, 96
Counterfeit coinage, 133–5, 137, 139, 151–4
Crank, applied to steam engine, 114; substitute for, 114
Crowley's works near Newcastle, 180
Crown mark assigned to Sheffield plate, 69
Cylinder boring mill, 85, 88, 92, 168

Danes visit Soho, 72
Dartmouth, Lord, supports Assay Bill, 69; letters to, 83, 135
Darwin, Dr Erasmus, friend of Boulton, 37, 208; learns of Watt's engine, 79; panegyric of engine, 126–7; description of Soho Mint, 157; his talking machine, 185; death, 189
Day, Thomas, 88, 208
Deceased wife's sister, marriage with, 34, 35
Denmark, mint for, 156, 207
Die sinking, 135, 146, 147, 153, 206
Dixon, Croat, 201
Dodding, William, 201
Dollars, Mexican, recoined, 155
Drawing office, 116, 171
Drawing schools, in Birmingham, 63
Droz, J. P., die sinker, 136, 137, 144, 146, 147, 158, 206
Dumergue, Charles, 208
Dundonald, Earl of, 120
Dutch visit Soho, 72
Duty on Cornish engines, 95

East India Company, coinage for, 136, 206; sheathing for, 160; reels for, 205, 206
Easton, James, 173
Economical Society of St Petersburgh, 199
Edgeworth, Richard Lovell, 35, 88
Eginton, Francis, mechanical paintings of, 104–7, 205

Elliott and Praed, bankers, 97
Ellis Greenberg collection of Boulton plate, 52, 53
Engine governor, 125, 126
Engines, how built, 90, 163, 164
Enticement of artisans, Act against, 49, 50
Etruria, Wedgwood's works at, 39, 47
Ewart, Peter, 168
Eyre, Lord Chief Justice, 174

Family tree, Boulton, 27
Farquaharson, James, 49
Feast, Soho Foundry rearing, 168, 169
Flaxman, John, 193, 198
Flour milling, engine applied to, 122–4; improvements in, 123
Fly-press, introduction of, 13; applied to coinage, 133
Foley, Richard, and the slitting mill, 11
Fordyce, James, failure of, 83
Fothergill, John, Boulton's partner, 43, 45, 46, 203, 204; unsuccessful, 109; partnership dissolved, 110; death, 111
Fothergill's family, 195, 200
Foulds, William, burglar, 184
Foundry, Soho, established, 167; rearing feast, 168–9; present day, 170; profitable, 172; Insurance Society, 181
Franklin, Benjamin, letter from, 36; advises Boulton on his steam engine, 75, 76
Freer, Rev. T. L., 192
Fuel economy, change from vegetable to mineral, 3, 4

Galton, Mary Ann, 165
Galton, Samuel, 88, 201
Garbett, Samuel, friend of Boulton, 38, 39, 88, 162, 208; supports Assay Office petition, 67; visited by Watt, 79, 203
Gas-lighting, Murdock's, 190; illumination of Soho by, 190
Gauge, Boulton's vacuum, 116
George III, audiences of, 56, 149, 197; loyal address to, 130; proposes to visit Boulton, 198
Germans visit Soho, 72
Gimblett, John, watch making by, 58, 64
Glass making in Staffordshire, 14, 100
Glass painting, Eginton's, 106
Goldsmiths Company, London, oppose Assay Bill, 67, 68
Goodwyn's engine, 204
Governor, steam engine, 125, 126
Great Haywood, 6, 17

Halifax, Lord, suspects Boulton, 49
Hall-marking silver plate, 63–70
Hammer, steam, 115
Hammered money, 133
Hancock, Joseph, and Sheffield plate, 52
Handsworth parish, Manufactory in, 41–4; map of, 43; rejoicings in, 69; Boulton's improvements in, 208
Handsworth Parish Church, Boulton buried in, 192, 193
Harper's Hill, Watt's residence, 92, 163
Harrison, Bill, 147
Henderson, Lt. Logan, 187
Herschel, Sir William, 188
Hockley brook, 42, 43, 75
Hodges, John, 201
Holiday, Boulton takes a, 119
Hornblower, Jabez, 173, 175
Hornblower, Jonathan, 95, 120, 173
Horse power, 124, 176, 178
Hospitality, Boulton's, 87, 186, 195, 208
Hôtel des Monnaies, French, 136
Hydraulic ram, 173

Illness, Boulton's, 162, 190
Illuminations at Soho, 190
Industrial art, school of design, 60, 63
Industry in Tudor times, 1, 2; changes in, 3; vertical integration of, 41
Insurance Society, Soho, 179–82
Invention, stimulus to, 129
Inventiveness, Boulton's, 29, 116–18, 140, 195, 203, 207
Inventiveness of Birmingham men, 21

Ireland, visit to, 119
Iron, extraction of, 3, 4; working of, 5, 9, 11, 14, 20, 21

Jary's engine, 98
Jewellery, steel, 46–7
Johnson, Samuel, visits Soho, 72

Keen, Mary, 201
Keir, James, friend of Boulton, 39, 208; member of Lunar Society, 88; career, 100; at Soho, 101, 207; his compound metal, 101–3, 161; his colliery, 104; partnership in copying press, 108; his memoir of Boulton, 70–2, 100, 107, 196, 197
Keir, Mrs, Boulton's present to, 100; death of, 191
Kinneil, Watt's engine at, 80, 82, 83, 90, 116
Kinneil House, 80, 119
Küchler, C. H., die sinker, 146, 147, 159

Labour, sub-division of, 41
Lamp, Argand's, 127–9
Lander, George, undertaker, 192, 193
"Lap" engine, 125, 179
Lawson, James, 147, 201
Legge, Heneage, 201
Leland's account of the Black Country, 9, 10
Length, coins to serve as standards of, 148, 149
Letter-copying machine, Watt's, 107–9, 167
Litigation, patent, 128, 129, 173–5
Liverpool, Lord, 138, 139
Luc, Frances de, 201
Luc, J. A. de, 129, 145
Lunar Society, 88, 165, 185–9

Maberley, John, lawsuit with, 174
MacGregor, James, father-in-law of Watt, 93
Magellan, J. H. de, 128
Manufacturers, Chamber of, 131
Manufactures fostered, 5
Medallic work, Boulton's, 145–7, 158, 205, 206
Medusa frigate sheathed, 161
Memoir of Boulton, Keir's, 70–2, 87, 100, 107, 196, 197
Memoir of Boulton, Watt's, 107, 146, 156, 194, 195, 203–8
Memorial medal, Boulton, 159, 162, 202
Mercantilism, 2
Mill, rolling, 12, 42, 45, 60, 153, 178, 207
Mint of 1750, 134
Mint, Royal, 134, 137, 150, 155, 156, 205
Mint, Soho, 137, 138, 142, 155, 157
Mints supplied to foreign countries, 156, 206, 207
Montague, Hon. Elizabeth, 54
Montgolfier's hydraulic ram, 173
Morton, Richard, and Sheffield plate, 52
Mosley, John, 201
Muntz's metal, 103
Murdock, William, engaged, 97, 98; opinion of Boulton, 147; boring mill, 168; returns from Cornwall, 170; invents gas lighting, 190
Murray, Matthew, 171
Museum, British, Boulton at, 55
Mynd, Mary, 166, 200

Nail making, 10–12
Navigations, transport by, 7
Nelson, Lord, 145, 158, 201
Newcomen's steam engine, 77, 86
Nozzles made at Soho, 91
Nunn, George, 200

"Old Bess" steam engine, 179
Organization of Soho Foundry, 172
Ormolu, manufacture, 54, 56–8, 60, 70, 71, 203, 205

Pack transport, 6, 16, 19
Packington, co. Staffs, estate of, 30, 35
Paintings, mechanical, 104–7
Panegyric of Soho, Darwin's, 182; of the steam engine, 126
Parallel motion, Watt's, 114, 115
Paris, engines in, 98, 99
Partnerships of Boulton, 43, 86, 108, 111, 209
Parys Mine, Anglesea, 131

Patents, of monopoly, 2, 21; Watt's, 80, 85, 108, 109, 114, 164, 167; Boulton's observations on, 129; Boulton's, 140, 173
Pedigree, Boulton, 27
Performance of engines, 178
Perrier Frères, engines for, 98, 99
Phillp, John, 147, 196
Pictures, copying, 104–7
Pierson, James, 201
Pig iron, production of, 5, 9, 14, 20
Pitt, William, 130
Podmore, Rev. C., 24 *n.*
Poldice, engine at, 95, 204
Poles visit Soho, 72
Polygraphic pictures, 106–7
Portland, Duke of, 130, 149, 150
Portraits of Boulton, 198
Premium, on pumping engine, 93, 95–6; on rotative engine, 124, 125
Premium payments from Cornwall, 112, 176
Press copying, 107–9
Priestley, Dr Joseph, 187–9
Pumping engines, 75–8, 90, 91, 95, 98, 113, 116

Queen Sophia, 56, 57, 145

Raids coiners' dens, Boulton, 152–4
Ram, hydraulic, 173
Rehe, Mr, 205, 207
Rennie, John, millwright, 122, 208
Research, endowment of, 188; Boulton's suggestions for, 161
Revolution, French, 166, 188–9
Richmond, Duke of, 65
Riots, in Birmingham, 144, 189; in Cornwall, 132
River navigations, 7, 8, 16
Roberts, Samuel, "silver thread" edge to Sheffield plate, 52
Robinson, Anne, 27, 34, 35, 119
Robinson, Mary, 27, 30, 34
Roebuck, John, friend of Boulton, 38–49; sinks coal pits, 76, 78, 83; takes share in Watt's patent, 79; offers licence to Boulton, 81; bought out by Boulton, 84
Rotative steam engine, 113–16, 123–6, 204
Royal Institution, Boulton a proprietor of, 199
Royal Society, Boulton Fellow of, 188, 199
Royalty, on pumping engines, 93, 95–6; on rotative engines, 124, 125
Royalty payments from Cornwall, 112, 176
Russia, Emperor of, 201, 206
Russia, Empress of, 58, 72
Russia, mint for, 156, 207

Salisbury, Rev. C. E., xi
Sandbrook, William, 201
Sardinians visit Soho, 72
Savile, Sir George, 69
Scale, John, 111, 209
Science Museum, models at, 122; "Lap" engine at, 125
Scissel, 159, 161
Scotland, visit to, 119
Scott, Sir Walter, opinion of Boulton, 184
Segregation of trades, 14, 15
Severn river as means of transport, 16, 20
Sheathing for ships, copper, 159–62
Sheffield, Assay Office established at, 65–70
Sheffield plate, Boulton's, 51–3, 70, 179
Shelburne, Lord, 61, 64
Shipbuilding, 7
Shuckburgh, Sir George, 149, 150
Sierra Leone Company, coinage for, 138
Silver plate, Boulton's, 53, 63–71, 179
Simcox, James, 200
Slitting-mill, 11, 12
Small, Dr William, friend of Boulton, 38, 208; his clock, 59; member of Lunar Society, 88; receives Watt, 203; death, 85
Smeaton, John, 93, 188
Smethwick, works at, 167
Smiles, Dr S., biographer of Boulton, ix
Snow Hill, workshops at, 25, 26, 29, 42, 46
Soho, name of, 44, 73
Soho Foundry, 167–73, 181, 207
Soho House, hospitality at, 72, 87; description of, 184, 185

Soho Insurance Society, 179–82
Soho Manufactory, beginning, 41–3; extensions, 49, 50, 75; resources of, 60, 179, 203; visitors to, 71–4; threatened mob attacks on, 145, 189; Insurance Society of, 179–82; panegyric of, 182; burglary at, 183, 184; first engine at, 207; closing of, 170
Soho Mint, establishment of, 138; description of, 142–4; work carried out by, 152, 155; Darwin's panegyric of, 157
Solander, Dr D. C., 49, 188
Sophia, Queen, audiences of, 56, 57, 145
Southern, John, engaged, 116; letter from, 164; abilities, 171
Spain, mint for, 156
Stage coach, 17, 18, 19
Stage waggon, 17
Standard of length, coins to afford, 148, 149
Standardization of engines, 115, 116
Steam power, applied to coining, 135, 138, 140–4, 157; applied to flour milling, 122–5
Steel, cementation, 12, 47, 48, 203
Steel jewellery, 46
Stour and Salwarpe Navigation, 7, 16
Stourport created, 17
Stratford-le-Bow, engine at, 93, 204
Sulphuric acid, Roebuck's, 38
Sun and planet gear, 114, 125
Sweden, Boulton enticed to, 49, 50
Swellingrebel Mrs, 111, 195

Tar, Dundonald's coal-, 120
Taxation, Boulton's views on, 130
Taylor, John, button maker, 100, 121
Taylor and Maxwell, 166
Theatre, Birmingham, 191
Thicknesse, Philip, insults Boulton, 36
Tingtang, engine for, 94, 96, 204
Tokens, tradesmen's, 134, 139, 140, 158
Torryburn, engine at, 90
Trade depression, 82, 83
Trade gilds, influence of, 20
Trafalgar medal, 145, 158
Transport, land, 6, 17, 20
Travelling expenses, Boulton's, 120
Tregurtha Downs, engine for, 94
Trust, early instance of, 132
Tube boiler, Boulton's, 117, 118
Tudor policy in industry, 1, 2

Uxbridge, Earl of, 157

Vacuum gauge, Boulton's, 116

Walker, Ann, 200
Walker, Zacchaeus, 46, 109, 200
Watt, Gregory, 166
Watt, James, portrait of, 39; improves steam engine, 76, 78; visits Soho and meets Boulton, 79; takes out condenser patent, 80; licence proposed, 81; arrives in Birmingham, 84; extension of patent, 85; Boulton begins partnership with him, 85, 86, 93, 112; member of Lunar Society, 88, 189; erects first engines, 90; general section of engine, 91; resides at Harper's Hill, 92; second marriage, 93; engines in Cornwall, 94–7, 99, 113, 120, 121, 204; invents copying press, 107–9; patents rotative engine, 114–16; his parallel motion, 114; concessions to adventurers, 121; invents governor, 125; Soho Foundry, 163–6, 171, 207; distribution of engines, 176–8; end of partnership, 179; travels, 191; estimate of Boulton, 197; witnesses his will, 200; memoir of Boulton, 107, 146, 156, 194, 203–8
Watt, James, junr., education of, 166; organizes Soho Foundry, 172; Boulton's help of, 194
Watt engines, distribution of, 176–7; prices of, 178
Wedgwood, Josiah, friend of Boulton, 39; cameos, 47; sale room of, 62; member of Lunar Society, 88, 188; opposes Pitt, 130
Weights, coins to serve as, 148, 150, 152, 155
Welfare work, 143
Weston, Ambrose, 201

Weston, James, 201
Westwood, Arthur, xi, 68 *n.*
Wheal Busy, engine for, 94, 96
Whitbread's engine, 204
Whitehurst, John, 88, 207
Wilkinson, John, portrait of, 39; boring mill of, 85, 88, 92, 163; engine for, 90; tokens for, 140, 158; receives young Watt, 166; dispute with his brother, 168
Will, Boulton's, 200, 201
Wilson, Thomas, agent in Cornwall, 97, 130, 144, 176
Withering, Dr William, 88, 119
Woodward, John, 201
Workmen's insurance at Soho, 179–82
Wyatt, Samuel, architect, 122, 184, 208
Wyke and Green's pedometer, 96

Zabarinsky, Prof. P. P., 199 *n.*

CAMBRIDGE: PRINTED BY
W. LEWIS, M.A.
AT THE UNIVERSITY PRESS